Negative Thermal Expansion Materials

David J. Fisher

In everyday life, minute thermally-induced elongations are essentially invisible to the naked eye; but even minute expansions can fatally degrade device processing and performance in – for example – the semiconductor industry. Materials which, astonishingly, contract upon heating offer the great advantage of being able to tune the overall thermal expansion of composite materials or to act as thermal-expansion compensators. The development of these negative thermal expansion materials has advanced rapidly during the past fifteen years, and a wide variety of materials of differing types has now been identified, as well as a number of intriguing mechanisms which help to avoid the apparent inviolable tendency of size to increase with temperature. The present work is the most up-to-date summary of the current range of negative thermal expansion materials and of the associated mechanisms.

Negative Thermal Expansion Materials

David J. Fisher

Published by **Materials Research Forum LLC**
Millersville, PA 17551, USA

Published as part of the book series
Materials Research Foundations
Volume 22 (2018)
ISSN 2471-8890 (Print)
ISSN 2471-8904 (Online)

Print ISBN 978-1-945291-48-7
ePDF ISBN 978-1-945291-49-4

Distributed worldwide by

Materials Research Forum LLC
105 Springdale Lane
Millersville, PA 17551
USA
http://www.mrforum.com

Printed in the United States of America
10 9 8 7 6 5 4 3 2 1

Table of Contents

Introduction

General

A positive coefficient of thermal expansion is the norm and, if asked for a counter-example, even most scientists would make only some vague reference to water or ice ... and then to the alloy, Invar; before recalling that the latter's expansion is essentially zero, rather than negative. Even in the case of the ice-water system it is only the behavior of the water which is anomalous in the sense of the present book; most of the familiar change in density is due to a phase-transformation. So familiar is the sight of ice floating on water, that some film directors seem to believe that that behaviour is normal and may routinely arrange 'special effects' that depict gold, say, floating on its melt.

Over about the past quarter-century there has been an 'ever-increasing interest in decrease', so to speak. That is, an interest in solid materials which exhibit the unusual property of contracting in volume upon heating, or which have a negative coefficient of thermal expansion along at least one crystallographic axis.

Negative thermal expansion materials offer enormous industrial possibilities because they can control the overall thermal expansion of composite products; thus preventing thermal stress-fracture, for example. Inorganic substances can now be obtained which have coefficients of linear thermal expansion that exceed -30 x 10^{-6}/K. These giant negative thermal expansions, sometimes called thermomiotic behavior, open up a new era of control of thermal expansion in composites. Although negative thermal expansion had been previously observed in some simple materials at low temperatures it was the realisation, in 1996, that some materials exhibit the phenomenon over very wide ranges of temperature that boosted the current interest in it. The initial discovery of materials which exhibit negative thermal expansion over a wide temperature range, combined with explication of the mechanisms behind this unusual property, will indeed make sure that the materials find use in various applications, although a number of obstacles have so far impeded their widespread implementation.

Special attention has been paid to families of materials having open-framework structures consisting of corner-sharing polyhedra that are relatively rigid in themselves but are flexible in rotation. One such polyhedron is $Zr(WO_4)_2$, where the zirconium can be replaced by hafnium and the tungsten can be replaced by molybdenum. Another basic unit is $A^{4+}P_2O_7$, where A is any element having a valance of +4 and both A and P can be partially replaced by other elements which respect the charge balance. In the case of the $Sc_2(MO_4)_3$ unit, the scandium can be replaced by yttrium, aluminium or lutetium. More complex units include $Zr_2P_2WO_{12}$ and $Ca_{1-x}Sr_xZr_4P_6O_{24}$, where strontium can be replaced by barium or magnesium. Some negative thermal expansion materials consist of a framework structure comprising relatively soft polyhedra linked by stiff cyanide bridges. These include $Zn(CN)_2$ and $Cd(CN)_2$. Finally there are the perovskite manganese nitrides where the negative expansion is produced by a magnetovolume effect. Also of interest are the quartz-like metal-organic framework materials, deuterium indium terephthalate and zinc isonicotinate. These exhibit anisotropic positive and negative thermal expansions. In the former material the negative-expansion response is uniaxial and occurs along the hexagonal crystal axis. In the later material positive expansion occurs along the hexagonal axis and negative expansion is found over the entire plane of perpendicular directions. This inversion of the mechanical response can be explained merely on geometrical grounds. A critical framework geometry can be identified that demarcates linear from areal negative-expansion behaviour. This analysis can be extended to other common metal-organic framework topologies. It suggests that framework geometry plays a pivotal role in determining the mechanical response of framework materials that exhibit an anisotropic response via hinging.

In general the negative-expansion materials involve strong coupling between other physical properties and the expansion. In both ferro-electrics and magnetics there is a common theme, in which the negative expansion is governed by either ferro-electric or magnetic order. Control of the expansion can be established via the coupling role played by ferro-electricity, magnetism, a change in electron configuration or the open nature of the framework, etc. Chemical modification is a particularly effective means for controlling thermal expansion.

Materials with low coefficients of thermal expansion are critically important to thin-film design of efficient bimorph actuators, and for stabilizing low-stiffness structures thermally. The first negative thermal expansion material, zirconium tungstate, has been prepared as a thin film and examined using a variable-curvature micro-mirror. The measured coefficients of various films are a function of stoichiometry and annealing conditions. Coefficients which are as low as -10×10^{-6}/K have been measured and have exhibited no hysteresis after several annealing cycles.

A negative expansion ceramic substrate and a combined athermal fiber Bragg grating component have been subjected to reliability tests, confirming that the component has enough durability to use it as an optical filter under conditions of damp, low temperatures, mechanical shock and vibration.

The use of negative thermal expansion particles as composite fillers is new and particle/matrix interface behaviour has not been greatly studied. In microchip packaging the composite may be constrained by the surroundings and be unable to expand during heating.

Thirty years after the Challenger disaster, engineers still wrestle with the fact that the coefficient of thermal expansion of rubber is more than an order of magnitude greater than that of steel and that an elastomer seal - compressed into its groove - may lose contact with the mating steel surface in cold weather; leading to potentially disastrous leaks. The volume fraction and shape of filler particles affect the thermoelastic properties and the sealing ability of elastomer composites. The contact pressure-drop due to freezing can be twice reduced in the case of spherical inclusions and the effect can be greater increased by changing the shape of filler particles. It is clear that filler particles which expand with decreasing temperature will have enormous beneficial effects upon safety.

Construction of a multi-layered system built from conventional materials with differing mechanical and thermal properties has been proposed. This system could exhibit predetermined values of thermal expansion; including negative ones. In order to maximise the negative thermal expansion there should then be a combination of thin layers of stiff materials having a high positive coefficient of thermal expansion, bonded to thicker layers of a soft material having a low coefficient of thermal expansion[1]. The Poisson ratio should also be as high as possible. A simple cylindrical structure with a stiff needle-like inclusion embedded a much softer matrix has been suggested to be a system with tunable thermal expansion[2]. With the correct combination of thermal and mechanical properties of the matrix and inclusion, such a system can be tailored so as to exhibit particular values of thermal expansion in the radial direction and also negative thermal expansion. Analytical expressions for the in-plane and out-of-plane coefficients of thermal expansion of laminates with isotropic laminae have been based upon conditions of zero net force on the laminate edges, equal displacement of the lamina edges and Hooke's law in three dimensions. Laminates consisting of alternating auxetic (positive coefficient of thermal expansion) and negative thermal expansion laminae (with positive Poisson ratio) exhibit a more negative effective coefficient than do laminates consisting of alternating conventional laminae (positive Poisson ratio and positive coefficient) with non-conventional laminae (auxetic and negative coefficient)[3]. Under certain circumstances, the magnitude of the temperature-dependent effective coefficient of thermal expansion tends to infinity. Using multi-material projection micro-stereolithography, lightweight lattices have been prepared which exhibit significant negative thermal expansion in three directions over a range of 170°. The negative expansion is due to the structural interaction of components having distinct thermal expansion coefficients[4]. The latter can be tuned over a large range by varying the thermal expansion coefficient difference between constituent beams.

One concept for reducing the thermal expansion of composites is to exploit the Poisson effect, Systems made from hard high-expansion needle, cylindrical or coin-shaped inclusions, embedded in a soft matrix having a high positive Poisson ratio and small coefficient of thermal expansion, can exhibit negative thermal expansion under certain conditions. In the case of coin-shaped inclusions, the negative expansion is greatest in the direction orthogonal to the surface of the coins[5]. In the case of needle-shaped inclusions, the negative expansion is greatest in the direction orthogonal to the length of the needles.

As well as preparing composites from conventional materials, a lively field of interest is the preparation of metamaterials where the structural level is microscopic rather than atomic. In fact, there is a close similarity between some of the atomic models used to explain negative expansion and the structures found in metamaterials. A novel method for obtaining two- and three-dimensional negative thermal expansion metamaterials via antichiral structures is to prepare two-dimensional metamaterial from unit cells that combine bi-material strips and antichiral structures. A three-dimensional metamaterial can be fabricated by means of multi-material three-dimensional printing. Experimental and simulation results all confirm the existence of isotropic negative thermal expansion properties. The effective coefficient of negative thermal expansion depends upon the difference between the thermal expansion coefficient of the component materials, as well as upon the circular node radius and the ligament length in the antichiral structures. The measured value of the linear negative thermal expansion coefficient of the three-dimensional sample is among the highest experimental values so far achieved[6]. This is an easy and practical approach to obtaining materials, of any scale, having tunable negative thermal expansions. A method has been found for designing metamaterials which can simultaneously exhibit a negative Poisson ratio and a negative thermal expansion[7]. These unusual properties depend mainly upon particular features of the micro- or nano-structural deformation when subjected to thermal or mechanical loads. Three-dimensional two-component polymer micro-lattices have been prepared by using grey-tone laser lithography. Displacement-vector field results reveal that the thermal expansion and resultant bending of bi-material beams leads to rotation of the chiral crosses arranged onto a three-dimensional chessboard pattern within one metamaterial unit cell[8]. The latter rotations can compensate for the expansion of the all-positive constituents, leading to an effective near-zero thermal length-expansion coefficient, or can over-compensate the expansion and lead to an effectively negative thermal length-expansion coefficient.

A triangular system made up of three beams welded together has been studied, in which the base of the triangular construct had a different thermal expansion and Young's modulus to those of the sides. The system was capable of exhibiting negative thermal expansion in one dimension when the base of the triangle had a larger thermal expansion coefficient than that of the sides. It could also exhibit negative compressibility[9]. The degree of linear thermal expansion depended upon the relative Young's moduli of the materials that made up the base and sides.

Anisotropic positive and negative thermal expansion behaviours in a cellular microstructure have been investigated in which the representative volume element of the structure consisted of thermally inextensible and thermally extendible rod elements. The degree of negative thermal expansion was greatly reduced when the central junction angles were rotatable and the rods were extendible[10]. The coefficient of thermal shearing depended upon the microstructural geometry and the change in temperature, but was independent of the coefficient of thermal expansion.

Material microstructures have been described in which the coefficient of thermal expansion is larger than that of either constituent. The thermal expansion can be large and positive, zero or large and negative. Three-dimensional lattices with void space exceed two-phase bounds but obey three-phase bounds. Lattices and normal materials tend to exhibit expansions which decrease with the modulus[11]. Two-phase composites which possess a negative stiffness phase exceed limits based upon a positive strain-energy density.

Unimode metamaterials made from rotating rigid triangles have been shown to exhibit positive Poisson ratios of huge magnitude for certain conformations. When the Poisson ratio in one loading direction is greater than unity, some systems exhibit negative linear compressibility along that direction: the system expands in this direction when hydrostatically compressed[12]. One structure which exhibits a negative Poisson ratio is the re-entrant triangle. A re-entrant triangular cellular structure composed of rigidly joined ribs of two different materials exhibits a negative thermal expansion which depends upon the inclination of the longer chevron strut, the dimensionless rod coefficient of thermal expansion and the ratio of the thermal expansion coefficients of the constituent struts ... but is independent of scale and temperature. The magnitude of the property becomes more negative at higher angles of the longer chevron strut with respect to the vertical, with higher ratios of thermal expansion of the base and chevron strut materials and with decreasing non-dimensional re-entrant base material thermal expansion coefficient[13]. The anisotropic and negative thermal expansion behaviour is stretch-dominated.

Elementary theory

In 1960, K.G.McNeill published[14] an amusing paper in which he predicted the values of a number of physical properties, among them the coefficient of thermal expansion, by using the simplest-possible assumptions and calculations. His methods unsurprisingly left no room for any possibility of negative thermal expansion in a typical crystal structure made up of atoms undergoing simple vibrational motions.

As Munn later pointed out, the thermal expansion of solids is the result of two effects. One of them is a strain-dependent entropy which produces a so-called thermal pressure that is the driving force for thermal expansion. In the case of rubber, the expansion is an entirely entropic phenomenon. This is why a party balloon puckers when it gets too close to a heat-source and why an elastic band becomes noticeably colder as one stretches it across ones hand.

Thermodynamic theory indeed suggests that substances which exhibit positive thermal expansion produce heat under compression while those exhibiting negative thermal expansion absorb heat. On the other hand, experiment shows that substances exhibiting negative thermal expansion can in fact produce heat. Attempts have been made to explain this paradox[15].

It has been shown that, for substances exhibiting positive thermal expansion and positive compressibility and for those exhibiting a negative thermal expansion and a negative compressibility[16],

$$\delta Q = dU + PdV$$

For substances exhibiting positive thermal expansion and negative compressibility, and for substances exhibiting negative thermal expansion and positive compressibility,

$$\delta Q = dU - PdV$$

These results help to treat processes which do not obey traditional thermodynamics[17].

In most materials thermal expansion is described by Grüneisen functions and the elastic response to this force. In cubic and isotropic solids, the thermal expansion is simply proportional to the product of the Grüneisen function and the bulk compressibility. The compressibility is obliged to be positive, for mechanical stability, and varies slightly and monotonically with temperature. Any unexpected behaviour of the thermal expansion, such as becoming negative or changing greatly with temperature, are then attributed to peculiarities of the Grüneisen function.

Among non-cubic solids (table 1), Munn[18] noted there appeared to be no cases of negative bulk thermal expansion although one of the principal coefficients of thermal expansion is often negative. Non-cubic symmetry also permits one linear compressibility coefficient to be negative even though the bulk compressibility is required to be positive. The sign of the thermal expansion is not governed by those of the Grüneisen functions or the linear compressibilities in the case of indium. Here the lattice is tetragonal and gives no indication of marked anisotropy but, above about 50K, the coefficient of thermal expansion nevertheless decreases steadily; becoming -7.7 x 10^{-6}/K above 200K.

Table 1 Thermal Expansion Coefficients of Some Axial Materials.

Material	Temperature (K)	Orientation	Coefficient (/K)
arsenic	283	perpendicular to axis	1.2 x 10^{-6}
arsenic	283	parallel to axis	41 x 10^{-6}
arsenic	30	perpendicular to axis	-2.0 x 10^{-6}
arsenic	30	parallel to axis	15 x 10^{-6}
graphite	1000	perpendicular to axis	0.8 x 10^{-6}
graphite	1000	parallel to axis	30 x 10^{-6}
graphite	300	perpendicular to axis	-1.2 x 10^{-6}
graphite	300	parallel to axis	28 x 10^{-6}
selenium	300	perpendicular to axis	69 x 10^{-6}
selenium	300	parallel to axis	-4.5 x 10^{-6}
tellurium	300	perpendicular to axis	30 x 10^{-6}
tellurium	300	parallel to axis	-2.3 x 10^{-6}
indium	284	perpendicular to axis	51 x 10^{-6}
indium	284	parallel to axis	-7.7 x 10^{-6}
indium	25	perpendicular to axis	4.8 x 10^{-6}
indium	25	parallel to axis	21 x 10^{-6}

The thermal expansion of silicon and titanium has been calculated by using density-functional perturbation theory. The phonon-mode contribution to thermal expansion has been analyzed and the negative thermal expansion is shown to be dominated by negative mode Grüneisen parameters at specific points on the Brillouin zone boundaries. The elastic Debye theory for negative thermal expansion is thus irrelevant here[19]. The anomalous behavior of these modes in titanium is unaffected by any electronic topological transition and instead arises from a complex interaction of the atomic displacements of the anomalous mode.

Both arsenic and graphite have layer structures, although the binding between layers is much weaker in hexagonal graphite than in trigonal arsenic. When the thermal expansion of arsenic was measured between 2 and 283K, the perpendicular component was -2.0 x 10^{-6}/K at low temperatures but became positive at higher temperatures; as in the cases of graphite, zinc, cadmium and white tin. In graphite, the value was -1.2 x 10^{-6}/K at room temperature. Both selenium and tellurium crystallize in a trigonal form made up of helical chains of atoms; the opposite extreme to layer structures. Their coefficients of thermal expansion are -4.5 x 10^{-6} and 2.3 x 10^{-6}/K, respectively.

Because it is the primordial example of such materials, the free energies of two low-pressure forms of ice have been calculated over a wide range of temperatures in order to explain their negative thermal expansions in the low-temperature regime. It is found possible to deduce the thermal expansion, at a given temperature and pressure, from intermolecular interaction potentials alone. The negative thermal expansion arises mainly from bending motions of hydrogen-bonded molecules.

It is clear that none of these elementary substances, nor indeed elementary analysis, will usher in a new field of materials science and it is fortunate therefore that intriguing new compounds possessing the desired property have been identified. The phenomenon has now been shown to occur in a range of solids, including complex metal oxides, polymers and zeolites. Increasingly quantitative studies have meanwhile focused on determining the mechanism of negative thermal expansion via high-pressure diffraction, local structure probes, inelastic neutron scattering and atomistic simulation.

A systematic EXAFS study has been performed on negative thermal expansion crystals having diamond-zincblende, cuprite or delafossite structures. Among the common features which appear, it is found that the nearest-neighbour bond expansion is always positive. Within each family of isostructural compounds, the stronger the lattice negative thermal expansion the greater the positive bond expansion[20]. A correlation was established between the negative expansion properties and the anisotropy of vibrations with respect to the bond direction. An EXAFS expression has been introduced for the analysis of positive or negative thermal expansion in which the EXAFS parameters are directly related to the interatomic potential[21]. The sign of one of the anharmonic potential parameters depends upon the sign of the thermal expansion coefficient.

Even though the coefficient of thermal expansion is an intrinsic thermophysical property of a material, measured values can depend upon sample preparation techniques and can exhibit hysteresis and scatter, perhaps due to microcracking, grain boundary separation, plastic deformation and slow phase transformations.

Various mechanisms have been identified, some involving a small number of phonons that are related to the rotation of rigid polyhedral groups of atoms and others involving large bands of phonons. In some cases, transverse acoustic modes make the main contribution to the negative expansion. The elasticity of these materials under pressure has also been studied and has revealed elastic softening. This so-called pressure-induced softening is closely linked to the negative expansion. Upon considering the lattice dynamics between bond strain and transverse vibrations, a common model has been proposed for ceramic and hybrid materials which exhibit negative thermal expansion. It is argued that pressure-induced softening, an increase in negative expansion with pressure and variations in negative expansion with temperature all arise naturally from the same mechanical model that explains negative thermal expansion itself[22].

There is interest moreover in the role played by intrinsic anharmonic interactions, and in making calculations of the potential energy wells for relevant phonons. Anharmonicity affects the response to pressure of the properties of the materials. Molecular dynamics simulations of cubic siliceous zeolites suggest that pressure-induced softening is likely to be a common feature of framework materials that exhibit negative thermal expansion[23]. By using a simple model of a negative thermal expansion system, a phenomenological expression was found for describing the temperature dependence of pressure-induced softening in negative thermal expansion structures[24].

Overall there is a multitude of effects which can lead to negative thermal expansion: at weak concentrations of iron-group transition ions in cubic crystals, the contribution to thermal expansion will be negative when the zero-field splitting energy is due to a second-order spin-orbit perturbation[25]. In some heavy fermion systems at low temperatures, within a two-component Fermi-liquid model, it may become negative if the Fermi-level falls on a region of steep increase in the density-of-states[26]. This also accounts for a correlation between the sign of thermal expansion and that of the thermopower. When the pressure equation of state of a classical many-body system with Gaussian pair interactions is examined over a wide density range, a simply-connected region of negative thermal expansion emerges[27]. It includes portions of the body-centered cubic crystal and liquid regions of the equilibrium phase diagram.

Framework structures

The fact that the currently most important class of negative thermal expansion materials is that of zirconium tungstate and its analogues has led to a lot of modelling of the relevant structures. The latter are those which consist of relatively stiff units, such as octahedra or tetrahedra, that are linked by shared atoms at their corners. Rigid rotation of the units often leads to a reduction in the volume or in that of some lattice constant (figure 1). It is a purely geometrical effect; rather like a child's 'spinner toy' in which a loop of non-elastic string is threaded through two closely-spaced holes in a disk. When the disk spins, the hands holding two ends of the loop are drawn together. This geometrical effect makes a negative contribution to the thermal expansion coefficient that is additional to the usual positive contribution arising from anharmonicity of the interatomic forces. The negative effect varies across the phonon spectrum; being greatest at low frequencies. The sign of the temperature coefficient can reverse above a soft-mode phase-transition.

The rigid unit mode model is a useful computational method for investigating correlations between transverse thermal motions of atoms involved in negative thermal expansion. Detailed calculations were made of 10 framework oxide structures that had been studied because of their negative or near-zero thermal expansions. The structures were essentially networks of polyhedra connected only by their corners. All of the evidence supported the importance of transverse motions of the atoms at the corners, but the negative expansion did not correlate well with the supposition of rigid unit modes for the polyhedra[28].The results denied the existence of any simple direct correlation between the presence or absence of rigid unit mode behaviour in a structure, and its negative expansion.

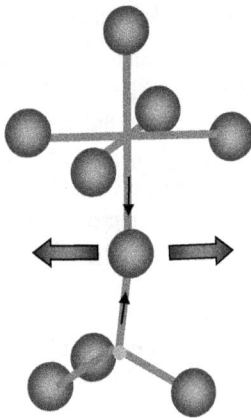

Figure 1 Oscillation of the atom linking octahedron and tetrahedron draws the groups together.

The negative thermal expansion of a space frame consisting of tetrahedral volume elements made up of two types of material was studied. Based upon these models, conditions for attaining a negative volumetric thermal strain and a negative coefficient of volumetric thermal expansion were established[29]. A trend was revealed in which the extent of negative expansion increases at lower apex-to-base rod length[30].

A topology-optimization method has been proposed for designing planar periodic frame structures that exhibit negative thermal expansion. If the beam section of each existing member is chosen from a finite set of many predetermined candidates, the topology optimization problem with multiple material phases can be formulated as a mixed-integer linear programming task and a global optimum solution can be readily found[31]. Because the method treats frame structures and local stress constraints, the optimum solution does not contain thin members or hinge-like regions.

There is some value in treating simplified systems which capture the essential features of frame structures, without having to handle all of the complexities of three-dimensional rotations. The analysis of a two-dimensional structurally-rigid construct made from rods of various materials, connected via hinges to form triangular units, shows that such a system can be made to exhibit negative thermal expansion along certain directions. They can even exhibit thermal expansions that are more positive than those of any of the component materials[32]. The end result is a multifunctional system possessing thermal properties that can be tailored for particular applications.

Existing rigid unit mode models, based upon rotating squares, have been generalized in order to treat systems made up of rectangular or rhombic rigid units. All of the systems exhibited negative thermal expansion to an extent which was governed by the shape and connectivity of the elemental rigid units. Some of these networks could potentially exhibit much greater negative thermal expansions as compared to those of networks of squares. In order to obtain optimum negative expansion properties, pencil-like rigid units were recommended rather than square ones[33]. The former permit larger pore sizes, and that promotes negative expansion.

As for the bonds between atoms, negative thermal expansion has been studied in model lattices having several atoms per cell and first- and second-nearest neighbor interaction involving the anharmonic Morse potential. By permitting all of the bond lengths to expand at different rates, it is found that negative thermal expansion is possible without requiring fully rigid units. Almost constant large-amplitude isotropic negative expansion is observed to occur up to the melting temperature in a molecular dynamics model of a ReO_3-like structure when rigidity of the octahedral units is almost entirely eliminated. Only slight negative expansion, changing to positive, is observed when the corner-linked octahedra are rigid, with flexible second-nearest neighbor bonds between the octahedra permitting easy rotation. Similar changes in thermal expansion behavior occur in the diamond lattice, with negative expansion when the second-nearest neighbor interactions are weak and positive thermal expansion when the second-nearest neighbor interactions are strong[34]. It is deduced that the only essential local conditions for negative expansion are atoms of low coordination number, together with very low energies for bond-angle changes relative to those for bond-stretching.

An isotropic interaction potential has been devised that gives rise to a negative thermal expansion behavior in equilibrium two- or three-dimensional many-particle systems over wide temperature and pressure ranges. Optimization procedures identify a potential that yields a strong negative thermal expansion effect; a key feature of the potential giving rise to this behavior being a softening of its basin of attraction. Such an anomalous behavior occurs in materials with directional interactions, such as zirconium tungstate[35]. Constant-pressure Monte Carlo simulations show that, as the temperature is increased, the system exhibits negative, zero and then positive thermal expansions before melting. Negative thermal expansion is suggested to originate from the existence of high-pressure low-volume phases of higher entropy in materials such as Fe_3Pt and ZrW_2O_8[36].

Thermal expansion of a classical chain with pair interactions, performing longitudinal and transverse vibrations, has been investigated. To a first approximation, the equation of state is of Mie–Grüneisen form and, for a Lennard-Jones type of potential, the Grüneisen parameter varies with deformation from minus infinity (zero stretching) to plus infinity (breakage point). At small deformations and finite thermal energies, the Mie–Grüneisen equation of state greatly overestimates the absolute value of the pressure. A more accurate equation implies that the thermal pressure in the unstretched chain is proportional to the square root of the thermal energy[37]. Near to the deformation corresponding to a zero Grüneisen parameter, the chain exhibits negative thermal expansion at low temperatures and positive thermal expansion at higher temperatures. A recent simple model assumes a single floppy mode for describing negative thermal expansion. An important feature is that the translational kinetic energy of the dilating bond network scales with the system size, predicting dynamic properties which differ qualitatively from those based upon harmonic models[38]. Marked differences with respect to conventional phonon dynamics and thermodynamics are seen in both the classical and quantum limits, including a marked enhancement of the negative coefficient of thermal expansion. Another model for materials exhibiting negative thermal expansion involves an arbitrary force between rigid units and has been applied to materials with a Lennard-Jones potential and to metal-organic framework materials with rigid units connected by organic linkages. The model can predict the temperature of the transition from negative to positive thermal expansion[39]. It is the same relationship as that for the temperature at which the glass transition begins to occur in many polymers.

Chapter 1

Metals

As it was an alloy (Guillaume's 'Invar') for which it was first realized that drastic changes in thermal expansion could be engineered, metals are thus an appropriate point at which to start a survey of the specific materials which exhibit *negative* thermal expansion. It will soon become apparent that metals and alloys are now very much in the minority. Nobel-laureate Guillaume, it will be recalled, described his Fe-36Ni zero-expansion alloy (actually 5 x 10^{-7} to 1 x 10^{-6}/K below 500K) in a classic paper[40]. Somewhat confusingly, the name of Invar has also been bestowed on negative thermal expansion alloys such as Fe$_3$Pt (-6 x 10^{-6} to -3 x 10^{-5}/K between 100 and 420K)[41,42].

A variety of alloys and composites have been used to develop negative thermal expansion metallic structures. In order to make parts from mixtures of alloys, novel methods are required as traditional brazing, welding or casting do not suffice. The direct metal deposition method, for instance, uses laser beams to melt coaxially injected metal. The present survey will concentrate however on the relevant physical properties.

AlLaCe

Measurements of the thermal expansion of the Kondo system, Ce$_x$La$_{1-x}$Al$_2$ ($0.07 \leq x \leq 1$), between 0.05 and 370K revealed a negative behaviour at sufficiently low temperatures[43]. This anomaly was explained in terms of a local Fermi-liquid model. Others have calculated the thermal expansion coefficient for a two-band Fermi liquid and have shown that the interband electronic transfer can make a large contribution. The calculations, as applied to the Kondo lattice, show that the thermal expansion coefficient can be negative for a certain range of values of the Kondo temperature and that a pseudo-gap is necessary[44]. Experimental results for Al$_3$Ce have been correctly interpreted using this model.

AuSrGe

Samples of Au$_3$SrGe, prepared by direct fusion and annealing, had a P4/nmm structure with a = 6.264Å, c = 5.5082Å and Z = 2 at room temperature. This structure, a re-proportioned $\sqrt{2}$ x $\sqrt{2}$ x 1 superstructure of CeMg$_2$Si$_2$ (P4/mmm), comprised checkerboard nets of corner-sharing bi-capped gold squares - or corner-shared octahedra

- in which the apical Au-Ge pairs in adjoining nets were strongly bonded in the c-direction. The Au_3SrGe exhibited marked uniaxial negative thermal expansion along the c-axis, with a coefficient of $-15.7 \times 10^{-6}/K$. The coefficient for the a- and b-axes was $21.6 \times 10^{-6}/K$[45]. The reticulated Au_5Ge octahedral layers expanded in the ab-plane upon heating, while the strong interlayer Au-Ge bonds remained fixed.

CoNd

Negative thermal expansion of the cobalt sub-network, of the order of 1% in volume, has been measured in amorphous $Nd_{1-x}Co_x$ (0.78 < x < 0.84) thin films at room temperature and at 10K. The expansion is anisotropic and is unrelated to a perpendicular magnetic anisotropy which is observed in all of the films. On the other hand, it is related to the method used to deposit them[46]. Because the magnetic moment of the cobalt sub-network is essentially independent of temperature, it has been proposed that the negative thermal expansion may be due to the mechanical response of the amorphous network to structural transformations at the neodymium sites.

Cesium

The equation of state of face-centered cubic cesium has been studied theoretically. The thermal expansion coefficient is predicted to be negative, at pressures above 3.5GPa, for all temperatures up to the end of the stability range of the face-centered cubic phase. The latter phase becomes unstable at about 4.3GPa, where a transverse phonon mode with a wave-vector along (110) becomes soft[47]. It appears that no isostructural transition occurs.

CuInCe

The coefficients of linear thermal expansion of the heavy-fermion compound Cu_2CeIn and its non-heavy analogue, Cu_2LaIn, were measured at 0.3 to 12K. In the heavy-fermion regime, the coefficient for Cu_2CeIn had a large positive value. Below 4K, a change to marked negative behaviour occurred. The negative contribution was associated with short-range antiferromagnetic order[48]. The temperature variation of expansion of Cu_2CeIn was close to that of Al_3Ce.

FeAlYMn

The $Y_2Al_3Fe_{14-x}Mn_x$ (x = 0 to 12) compounds have an hexagonal Th_2Ni_{17}-type structure and the unit-cell volume first increases slowly with increasing x before increasing rapidly with further increments in x. This implies that a positive spontaneous volume magnetostriction occurs in the magnetic state of these compounds. Between 185 and

200K the coefficient of thermal expansion is -75 x 10^{-6}/K[49]. The Curie temperature and the saturation magnetization of the compounds fall rapidly with increasing x[50].

FeGdCr

The $Gd_2Fe_{16.5}Cr_{0.5}$ compound, when annealed at 1243C, has an hexagonal Th_2Ni_{17}-type structure. A chromium atom which replaces an iron atom can sensibly increase the Curie temperature. An anisotropic anomalous thermal expansion was observed when in the magnetic state. Along the c-axis, the average linear thermal expansion coefficient was -2.79 x 10^{-6}/K between 294 and 472K and -30.9 x 10^{-6}/K between 472 and 592K. Along the a-axis, the average linear thermal expansion coefficient was 9.22 x 10^{-6}/K between 294 and 552K and -14.1 x 10^{-6}/K between 552 and 592K[51]. Between 472 and 592K, the average volume thermal expansion coefficient was -21.4 x 10^{-6}/K.

FeHfTa

A competition between ferromagnetic and antiferromagnetic phases on frustrated lattices in the hexagonal Laves phase compound, $Fe_2Hf_{0.86}Ta_{0.14}$, means that, at 325K, it orders into a 120° frustrated antiferromagnetic state with a much-reduced magnetic moment. There is also a simultaneous in-plane lattice contraction. Upon further cooling, accumulated distortion destabilizes the frustrated structure and the latter is completely relieved at 255K by a first-order transition to the ferromagnetic state[52]. At this point, there is an enormous volumetric thermal expansion of -123 x 10^{-6}/K.

FeLaSi

Compounds based upon $La(Fe,Si)_{13}$ are familiar magnetocaloric materials which exhibit marked negative thermal expansion around the Curie temperature. They have not been much considered for industrial application, generally because of the limited – circa 100C - temperature range of the phenomenon. When $LaFe_{11.2}Si_{1.8}$ alloy was annealed (1200C, 5h) it contained some α-Fe phase, while samples which were annealed at 1050C for 15 days or at 1100C for 2 days comprised a single $La(Fe,Si)_{13}$ phase having the $NaZn_{13}$ structure. The degree of negative thermal expansion was a linear function of the $La(Fe,Si)_{13}$ content[53]. The thermal expansion coefficient of samples annealed at 1050C for 15 days was -39.31 x 10^{-6}/K. Samples of $NaZn_{13}$-type $LaFe_{13-x}Si_x$ and $LaFe_{11.5-x}Co_xSi_{1.5}$ were prepared, and the sharp volume change of $La(Fe,Si)_{13}$-based compounds was made continuous. Upon increasing the amount of cobalt dopant in $LaFe_{11.5-x}Co_xSi_{1.5}$, the negative thermal expansion shifted to higher temperatures and the operating range of negative expansion broadened[54]. The linear thermal expansion coefficient of $LaFe_{10.5}Co_{1.0}Si_{1.5}$ attained -26.1 x 10^{-6}/K, with an operating range of 240 to 350K.

The operating range of negative expansion of $LaFe_{13-x}Si_x$ can be significantly broadened by adjusting the Fe-Fe magnetic exchange coupling as x varies from 2.8 to 3.1. The range of $LaFe_{10.1}Si_{2.9}$, for example, is extended to 220K. The coefficients of thermal expansion of $LaFe_{10.0}Si_{3.0}$ and $LaFe_{9.9}Si_{3.1}$ are homogeneous over a negative-expansion range of about 200K[55]. The tetragonal phase is gradually introduced into the cubic phase as the silicon content increases; thus modifying the Fe-Fe interatomic distance. When cubic $NaZn_{13}$-type $LaFe_{11.2}Al_{1.8-x}Si_x$ (x = 0.2, 0.3, 0.4, 0.5) compounds were prepared using arc-melting methods, the negative-expansion behavior was markedly enhanced by replacing aluminium with silicon, and the operating range shifted towards higher temperatures. The $LaFe_{11.2}Al_{1.8-x}Si_x$ retained the cubic $NaZn_{13}$-type structure as the temperature ranged from 20 to 270K; including the temperature region where negative thermal expansion occurred[56]. The improvement in behavior was again attributed to enhancement of the Fe-Fe magnetic exchange interactions with dopant silicon atoms. The effect of manganese doping upon $NaZn_{13}$-type $La(Fe,Si)_{13}$-based compounds was that the negative thermal expansion temperature-range shifted towards lower temperatures. This was attributed to a decrease in the Curie temperature with increasing manganese addition[57]. The average coefficient of thermal expansion around the Curie point changes slightly with increasing manganese content in $LaFe_{11.5-x}Mn_xSi_{1.5}$. Another idea is to replace the lanthanum with praseodymium. A consequent increase in the absolute value of the average coefficient of thermal expansion between 200 and 300K is attributed to an increase in spontaneous magnetization with increasing praseodymium content[58]. For example, the average coefficient of thermal expansion of $La_{1-x}Pr_xFe_{10.7}Co_{0.8}Si_{1.5}$, with x = 0.5, attains -38.5 x 10^{-6}/K between 200 and 300K; 18.5% greater than that for x = 0. The negative thermal expansion operating range of $LaFe_{10.0}Si_{3.0}$ broadened significantly upon introducing interstitial hydrogen atoms[59]. An increase in the ferromagnetic-to-paramagnetic transition temperature, due to stronger Fe-Fe magnetic exchange interactions, again contributed to a broadened negative-expansion operating range in $LaFe_{10.0}Si_{3.0}$ hydride.

FeMnErAl

The compound, $Fe_{12}Mn_4Er_2Al$, has been investigated by means of X-ray diffraction and magnetic measurements. The results indicate that the compound is single-phase, with a Th_2Ni_{17}-type structure, and its Curie temperature is about 230K[60]. X-ray diffraction data for 160 to 652K show that negative thermal expansion occurs at 236 to 259K, with an average coefficient of -20.3 x 10^{-6}/K.

FeNi

Objects with a fine-scale bi-metallic beam structure can exhibit negative thermal expansion. Such a thermoelastic metamaterial has been created via the co-extrusion of iron and nickel oxides, followed by reductive sintering[61]. A 162-unit sample made from Fe-60Ni and Fe-36Ni alloys had a thermal expansion coefficient of -3.0 x 10^{-6}/K.

FeTbCr

The compound, $Tb_2Fe_{16}Cr$, has an hexagonal Th_2Ni_{17}-type structure. Negative thermal expansion was observed between 292 and 556K[62]. The coefficient of the average thermal expansion was -1.53 x 10^{-6}/K between 293 and 500K, and -153 x 10^{-6}/K between 500 and 522K[63]. Magnetization measurements showed that a chromium atom, upon replacing an iron atom, increased the Curie temperature of Tb_2Fe_{17}.

FeTmSiCr

The $Tm_2Fe_{15}SiCr$ compound has an hexagonal Th_2Ni_{17}-type structure in which one chromium and one silicon atom, in replacing two iron atoms, can sensibly increase the Curie temperature. An anisotropic anomalous thermal expansion was observed when in the magnetic state. Along the c-axis, the average linear thermal expansion coefficient was -20.4 x 10^{-6}/K between 303 and 460K. Along the a-axis, the average linear coefficient was -80.9 x 10^{-6}/K between 370 and 410K[64]. Between 370 and 450K, the average volume thermal expansion coefficient was -20.8 x 10^{-6}/K.

$GdPd_3B_{0.25}C_{0.75}$

A study of polycrystalline samples of this metallic ordered cubic compound suggested that the negative thermal expansion which is observed does not arise from valence or magnetic instability of lattice ions; which is the general case for such metallic compounds[65]. It has been proposed instead that it is transverse vibrations at low temperatures, arising from site anisotropy, which induce the lattice contraction and thus results in isotropic negative expansion.

MnCoGe

Alloys of the form, MnM'X (M' = Co, Ni; X = Ge, Si, etc.), usually undergo a large volumetric change during martensitic transformation. This offers the opportunity of developing new negative thermal expansion materials, if the temperature range of the negative expansion can be extended. A huge negative thermal expansion is observed in finely powdered $Mn_{0.98}CoGe$, prepared by repeated thermal cycling or ball milling. In powders which have undergone 10 thermal cycles, the range is from 309 to 399K[66], and the thermal expansion coefficient is about -141 x 10^{-6}/K; one of the highest negative values observed. Compounds based upon MnCoGe exhibit giant negative thermal expansions during the martensitic transition from a Ni_2In-type hexagonal structure to a

5

TiNiSi-type orthorhombic structure. The expansion of the unit-cell volume can be as much as 3.9%, but these materials are not generally considered to be suitable negative thermal expansion materials because of the limited temperature range of the phase transition. In addition, as-prepared MnCoGe-based compounds tend to collapse into powder form. The use of 3 to 4% of epoxy can however introduce residual stress into bonded samples and thereby achieve broadening of the structural transition. As a result, the average coefficient of $MnCo_{0.98}Cr_{0.02}Ge$ can be -51.5 x 10^{-6}/K, with an operating range as wide as 210K[67]. Between 250 and 305K, the coefficient of -119 x 10^{-6}/K is almost independent of temperature.

MnCr

Body-centered cubic $Cr_{1-x}Mn_x$ alloys (x = 0.4 to 0.66) were prepared by high-temperature quenching. A large negative thermal expansion was observed at low temperatures for x = 0.6. Assuming the existence of two electronic states of manganese, having different magnetic moments and atomic volumes, this behaviour was explained in connection with variations in the hyperfine field distribution at ^{55}Mn nuclei[68].

MnFeNiGe

Detailed study of $Mn_{1-x}Fe_xNiGe$ (x = 0, 0.06, 0.11) reveals the occurrence of successive first-order transitions: a martensitic transition and an antiferromagnetic-ferromagnetic transition. A transition such as the latter one, occurring in the martensitic state, has rarely been observed. Other data confirmed its first-order magneto-elastic nature. The isothermal entropy change for a field change of 30kOe attained -25.8J/kgK at 203K for x = 0.11. The iron-doped samples exhibited a giant negative thermal expansion and near-zero thermal expansion in different temperature ranges[69]. The average thermal expansion coefficient for $Mn_{0.94}Fe_{0.06}NiGe$ attained -60.7 x 10^{-6}/K between 231 and 338K, and 0.6 x 10^{-6}/K between 175 and 231K.

NiMnGa

Polycrystalline $Ni_{55.5}Mn_{19.5}Ga_{25}$ alloy possesses an unmodulated martensitic structure ($L1_0$) at room temperature and undergoes a first-order magneto-structural transformation from paramagnetic austenite to ferromagnetic martensite upon cooling, leading to a refrigerant capacity of 55J/kgK for a magnetic field change of $3T^{70}$. Unlike the large negative thermal expansion caused by magneto-structural transformation, such an effect in martensite is not accompanied by any thermal hysteresis. The coefficient of thermal expansion is over 8 x 10^{-6}/K between 250 and 50K.

NiTi

The effect of heat treatment upon the negative thermal expansion of NiTi was investigated, showing that the negative value for aged (360C, 130h) alloy increases to -87.0 x 10^{-6}/K when the stress is increased up to 250MPa. It decreases with further increases in stress. Upon increasing the aging temperature or lengthening the aging time at higher temperatures, the negative expansion behavior worsened[71]. It is deduced that the internal stress field caused by aligned particles of Ti_3Ni_4 precipitates controls phase transformation, resulting in the negative thermal expansion behavior being due to both the volume change and the two-way shape-memory effect. Pre-strained TiNi wire exhibits an unusual turning-point, in its strain-versus-temperature curve, during the thermal cycle following one incomplete reverse transformation. The turning-point is closely related to the previous arrest temperature and the corresponding arrest strain; regarded as a memory of the arrest strain-temperature point[72]. This has been termed a negative thermal expansion arrest-point memory-effect.

By controlling the cold-rolling thickness reduction and heat-treatment temperature, the coefficient of thermal expansion can be varied from 21 x 10^{-6}/K to -11 x 10^{-6}/K due to the in-plane anisotropy of the thermal expansion. A very small coefficient of -0.53 x 10^{-6}/K (essentially Invar-like) between 353 and 123K is obtained at an angle of 33.5° to the rolling direction of severely cold-rolled sheet. The associated microstructure consists of a mixture of B2 nano-grains and retained or residual deformation-induced martensite. The anomalous thermal expansion anisotropy is due to the intrinsic anisotropic thermal expansion of the residual martensite. Invar-like behavior results from a cancellation of the positive thermal expansion of austenite by the negative thermal expansion of the residual martensite at 33.5° to the rolling direction[73]. A simple rule-of-mixtures model describes the anomalous in-plane thermal expansion anisotropy.

Nickel titanium nevertheless generally has a large positive coefficient of thermal expansion. On the other hand, $Zr_2P_2WO_{12}$ is a negative thermal expansion material. Powders of the latter were prepared using hydrothermal methods and TiNi alloy, with and without added $Zr_2P_2WO_{12}$, was compacted by hot-pressing sintering (1100C, 3h). The resultant ceramics had a coefficient of thermal expansion of about -2.012 x 10^{-6}/K. The $Zr_2P_2WO_{12}$ consisted of relatively homogeneous particles with a size of about 360nm[74]. With increasing $Zr_2P_2WO_{12}$ content, the relative density of the composite materials decreased and the hardness increased. The $Zr_2P_2WO_{12}$ itself can be made from ZrO_2, WO_3 and P_2O_5 via two different reaction routes: one of them is direct reaction, while another is by way of the intermediate phase, ZrP_2O_7. The reaction route is governed by the drying and sintering temperatures[75]. By controlling the reaction route, $Zr_2P_2WO_{12}$ can

7

be created within minutes. The as-synthesized specimens also have much smaller and uniformly distributed grains than do those made using conventional methods.

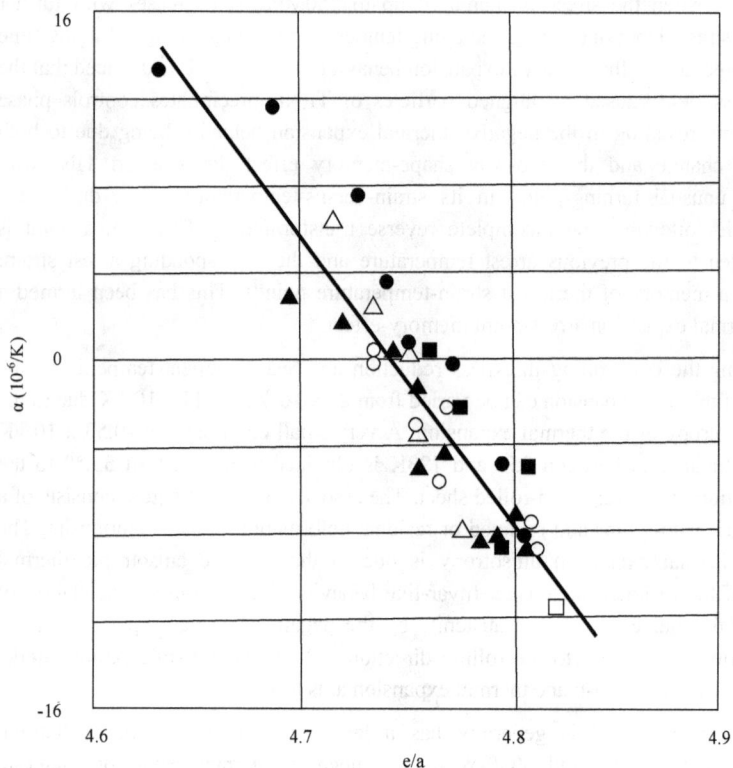

Figure 2 Coefficient of Linear Expansion of Plutonium Alloys.
Filled circles: Al, filled triangles: Ce, filled squares: In,
open squares: U, open circles: Zn, open triangles: Zr

Platinum

The thermal behavior of platinum nanoclusters supported on γ-Al_2O_3 was studied using scanning transmission electron microscopy and X-ray absorption spectroscopy. This revealed that the material had a very non-bulk nature. In an inert atmosphere the subnanometer Pt/γ-Al$_2$O$_3$ clusters underwent marked relaxation of the metal-metal bond distances and exhibited negative thermal expansion, with an average linear coefficient of -24 x 10^{-6}/K. Platinum nanoclusters which were supported on carbon-black did not exhibit negative thermal expansion[76]. These results highlight the significant role that metal-adsorbate interactions play in controlling the structural dynamics of supported nanoscale metal clusters.

Plutonium

The negative thermal expansion of δ-plutonium becomes less and less negative, and finally becomes positive upon adding increasing amounts of aluminium, zinc, zirconium, indium or cerium (figure 2). The change in expansion is related to the electron concentration, and the line in the figure has the equation,

$$\alpha = 6.3456 \times 10^{-4} - 1.3383 \times 10^{-4}(e/a)$$

where e/a is the electron/atom ratio. The expansion is zero at a critical electron concentration of 4.74; compared to the value of 4.82 for pure δ-plutonium[77]. It is positive for alloys whose electron concentration is less than 4.74.

Praseodymium

High-temperature angle-dispersive synchrotron radiation diffraction studies have revealed that a double hexagonal close-packed to face-centered cubic transformation occurs martensitically in praseodymium between 575 and 1035K. The high-temperature face-centered cubic phase exhibits negative thermal expansion between 600 and 800K[78]. This is attributed to 4f-electron delocalization.

TmFeCr

Compounds of the form, Tm$_2$Fe$_{17-x}$Cr$_x$ (x = 0, 0.5, 1.0, 1.5, 2.0), have an hexagonal structure and exhibit negative thermal expansion near to the Curie temperature, with a strong and anisotropic spontaneous magnetostriction in the magnetic state. The coefficient of thermal expansion is -9 x 10^{-6}/K between 340 and 380K[79]. The substitution of chromium for iron can greatly increase the Curie temperature and the easy-axis magnetocrystalline anisotropy. The Curie temperature of Tm$_2$Fe$_{16}$Cr is about 160K higher than that of the base compound, Tm$_2$Fe$_{17}$. The saturation magnetization of Tm$_2$Fe$_{17-x}$Cr$_x$ decreases with increasing chromium content.

Chapter 2

Molybdates

High-pressure powder X-ray diffraction studies of AMo_3O_{12} (A = Al_2, Fe_2, FeAl, AlGa) between 6 and 7GPa show that all of the materials have a monoclinic structure under ambient conditions and exhibit a similar phase transition behavior upon compression. The initial isotropic compressibility first becomes anisotropic, and is followed by a small distinct drop in cell volume. These patterns could be described by a distorted variant of the ambient pressure polymorph. At higher pressures, a distinct high-pressure phase forms. All of the materials have the same high-pressure phase. All of the changes are reversible upon decompression, apart from some hysteresis[80]. The transition pressures for all of the phase changes increase with decreasing radius of the A-site cation. High-temperature X-ray diffraction and dilatometry of $A_2Mo_3O_{12}$ (A = Y, Er, Yb, Lu) shows that the rare-earth molybdates are isostructural with the corresponding rare-earth tungstates. Those having an orthorhombic structure are highly hygroscopic and exhibit negative thermal expansion following the complete removal of water molecules[81]. Axial thermal expansion data show that, as the A-cation decreases in size, the thermal expansion coefficient for the b-axis and the linear thermal expansion coefficient become less negative. Materials of the form, $A_2M_3O_{12}$ (A = Al, Sc, Fe, In, Ga, Y; M = Mo, W), were prepared using a non-hydrolytic sol-gel process. Unlike previous attempts, in which amorphous materials resulted which then had to be crystallised by heating to between 500 and 700C, several compounds could be obtained at temperatures which were compatible with low-temperature processing[82]. Under some reaction conditions, crystalline $Sc_2Mo_3O_{12}$, $Fe_2Mo_3O_{12}$ and $In_2Mo_3O_{12}$ could be obtained at temperatures as low as 230C.

$Al_2(MoO_4)_3$

The thermal expansion behavior of $Al_2(MoO_4)_3$ was studied in still air, between 298 and 1073K, by means of thermodilatometry. It first exhibited a positive thermal expansion, followed by a phase transition and then a negative thermal expansion. It had a monoclinic structure at room temperature[83]. The thermal expansion coefficient was -2.83 x 10^{-6}/K.

$Al_2W_2MoO_{12}$

The nanostructured $Sc_2W_3O_{12}$-type negative thermal expansion material, $Al_2W_2MoO_{12}$, has been prepared by using the stearic acid method. Nanostructured $Al_2W_2MoO_{12}$ powders with an average size of about 10nm were obtained first, and then rearranged to form larger rods by prolonging the heat-treatment time[84]. The resultant nanoparticles exhibited negative thermal expansion between 20 and 600C.

$Cr_2(MoO_4)_3$

Powder samples were prepared by liquid-phase reaction precursor sintering method. *In situ* X-ray diffraction data, obtained between 298 and 1073K, revealed an essentially linear increase in cell volume as a function of temperature. The intrinsic linear coefficient of thermal expansion corresponding to these data was 1.612×10^{-6}/K. The thermal expansion behavior in still air between 298 and 1073K exhibited an initial positive thermal expansion, followed by a phase transition and then negative thermal expansion[85]. The thermal expansion coefficient here was -9.282×10^{-6}/K. The thermal expansion behavior of $Cr_2(MoO_4)_3$ was studied in still air, between 298 and 1073K, by means of thermodilatometry. It first exhibited a positive thermal expansion, followed by a phase transition and then a negative thermal expansion[86]. It had a monoclinic structure at room temperature. The thermal expansion coefficient was -9.39×10^{-6}/K.

$Fe_2(MoO_4)_3$

Monoclinic $Fe_2(MoO_4)_3$ was synthesized by using the hydrothermal method. A reversible phase transition between the low-temperature monoclinic and high-temperature orthorhombic phases occurs at about 510C. Between 25C and 400C, the a-, b- and c-axes of the monoclinic phase gradually expand. Between 530C and 710C, the orthorhombic phase exhibits a negative thermal expansion in which the b- and c-axes gradually contract but the a-axis first contracts and then expands slightly. Calculations indicate that the Mo-O bonds are much stronger than the Fe-O bonds and that the MoO_4 tetrahedra are more rigid than FeO_6 octahedra. Calculation of the Grüneisen parameters for phonon branches at the Γ point suggests that the optical branch with the lowest vibration frequency has the largest negative Grüneisen parameter. Analysis of the vibrational behavior of the atoms shows that oxygen atoms have differing vibrational eigenvectors to those of iron and molybdenum atoms, and greater amplitudes[87]. It is concluded that transverse vibration of the oxygen bridging atom between the MoO_4 tetrahedron and FeO_6 octahedron, soft distortion of the FeO_6 octahedra and rigid rotation of the MoO_4 tetrahedra together lead to the negative thermal expansion of $Fe_2(MoO_4)_3$. Samples were prepared from aqueous solutions of ferrous nitrate and ammonium molybdate by using a wet chemical method and then calcined (700C, 2h) to give a crystalline product with a monoclinic unit cell. A monoclinic-to-orthorhombic phase transition occurred at about 500C[88]. The $Fe_2Mo_3O_{12}$ exhibited a macro thermal expansion of -13.43 x 10^{-6}/K and a micro thermal expansion of -21.77 x 10^{-6}/K between 500 and 800C. The thermal expansion behavior of $Fe_2(MoO_4)_3$ was studied in still air, between 298 and 1073K, by means of thermodilatometry. It first exhibited a positive thermal expansion, followed by a phase transition and then a negative thermal expansion[89]. It had a monoclinic structure at room temperature. The thermal expansion coefficient was -14.82 x 10^{-6}/K.

$In_2(MoO_4)_3$

Orthorhombic $In_2Mo_3O_{12}$ has a low negative linear coefficient of thermal expansion, of -1.85 x 10^{-6}/K, and this is directly dependent upon the inherent volume distortion parameter of InO_6. This confirms a proposed relationship between the linear coefficient of thermal expansion in $A_2M_3O_{12}$ compounds, and the degree of distortion level of AO_6 polyhedra. With increasing inherent distortion of the AO_6 polyhedra, the linear coefficient of thermal expansion becomes more negative[90]. Another important feature of AO_6 polyhedra such as InO_6 is that their distortion increases with temperature. The orthorhombic $In_2Mo_3O_{12}$ is stable between 370 and 760C.

$Lu_2(MoO_4)_3$

The compound, $Lu_2Mo_3O_{12}$, was prepared by solid-state reaction, giving an orthorhombic structure which exhibited negative thermal expansion between 200 and 800C. The a- and c-axes exhibited a stronger contraction within that temperature range, while the b-axis slightly expanded between 200 and 300C and then contracted between 300 and 800C[91]. The linear thermal expansion coefficient was -5.67 x 10^{-6}/K.

$Sc_2(MoO_4)_3$

Thermal expansion of, and phase transitions in, this material were investigated between 4 and 300K by powder neutron diffraction and between 300 and 1053K by dilatometry. Below 178K, it has a monoclinic structure, $P2_1/a$, with a = 16.22715, b = 9.58051, c = 18.9208Å and β = 125.3988°. The coefficient of thermal expansion was 21.9 x 10^{-6}/K between 4 and 170K and there was significant anisotropy, with the monoclinic b-axis having a negative expansion coefficient between 4 and 86K. Above 180K, the material has an orthorhombic $Sc_2(WO_4)_3$-type structure and a coefficient of thermal expansion of -6.3 x 10^{-6}/K between 180 and 300K. Structural changes at the monoclinic-to-orthorhombic phase transition were closely related to a contraction of the orthorhombic phase[92]. Dilatometry measurements show that the negative thermal expansion behaviour continues up to 1053K. When samples of $Sc_2Mo_3O_{12}$ were prepared using solid-state methods and were examined by means of powder X-ray diffraction, scanning electron microscopy, thermogravimetric-differential thermal analysis and thermo-mechanical analysis[93]. The coefficient of thermal expansion, as determined using X-ray diffraction and thermo-mechanical analysis, was -6.25 x 10^{-6} or -5.62 x 10^{-6}/K, respectively. When $Sc_2(MoO_4)_3$ was prepared by the solid-state reaction of scandium oxide and molybdenum trioxide between 700 and 1100C, material of high purity and with an orthorhombic structure was obtained. The decomposition temperature was about 1116C, and the thermal expansion was apparently affected by the sintering temperature[94]. The average macroscopic thermal expansion coefficient changed from -4.49 x 10^{-6} to -5.83 x 10^{-6}/K between room temperature and 600C.

The behaviour of the solid solutions, $Sc_{2-x}Cr_xMo_3O_{12}$ (0 to 2), was investigated. At room temperature, samples with x = 0.7 and x = 0.8 had orthorhombic and monoclinic structures, respectively. A phase transition from monoclinic to orthorhombic occurred in $Sc_{0.5}Cr_{1.5}Mo_3O_{12}$ at 203.66C. The linear thermal expansion coefficient of the orthorhombic phase varied from 2.334 x 10^{-6} to 0.993 x 10^{-6}/K as x increased from 0.0 to 1.5[95]. Near-zero linear thermal expansion coefficients of 0.512 x 10^{-6} and 0.466 x 10^{-6}/K were observed for compounds with x = 0.5 and x = 0.7, respectively.

Raman spectra for $Sc_2(MoO_4)_3$ were obtained under pressures ranging from 0 to 3GPa. Lattice dynamics calculations were performed for the orthorhombic phase, using a rigid-ion model to calculate the phonon frequencies at the Brillouin zone center. The calculated results agreed qualitatively with the Raman spectra observed at room temperature under ambient pressure. A higher coupling among the modes of $Al_2(WO_4)_3$, as compared with $Sc_2(MoO_4)_3$, was observed. The 362 and 433/cm modes of $Sc_2(MoO_4)_3$ and $Al_2(WO_4)_3$, respectively, exhibited unusually high $\partial\omega/\partial P$ values[96]. The presence of these unstable modes in both materials was suggested to be related to the mechanism of the phase transition from the monoclinic high-pressure phase to an unknown phase of lower symmetry.

$Y_2(MoO_4)_3$

When powder samples of $Y_2Mo_3O_{12}$ were obtained by liquid-phase co-precipitation, followed by heat treatment (750C, 6h), the thermal expansion coefficient was -5.943 x 10^{-6}/K[97]. The individual coefficients for the a-axis, b-axis, c-axis and volume were -2.147 x 10^{-6}, -6.293 x 10^{-6}, -9.456 x 10^{-6} and -17.83 x 10^{-6}, respectively. The Pbcn orthorhombic phase of $Y_2Mo_3O_{12}$, isostructural to $Fe_2Mo_3O_{12}$ and consisting of vertex-sharing YO_6 and MoO_4 polyhedra, has been examined using high-resolution X-ray powder diffraction at 10 to 450K, heat-capacity determinations at 2 to 390K and differential scanning calorimetry at 103 to 673K. No phase transition was found within these temperature ranges. The overall thermal expansion was negative, and the average linear thermal expansion coefficient was an average of -9.02 x 10^{-6}/K between 20 and 450K and -12.6 x 10^{-6}/K between 130 and 900C[98]. The YO_6 octahedra and MoO_4 tetrahedra became increasingly distorted with increasing temperature. A so-called inherent volume distortion parameter was introduced in order to evaluate the polyhedral distortion quantitatively, and this parameter was strongly related to the linear coefficient of thermal expansion for various members of the $A_2M_3O_{12}$ family[99]. The negative thermal expansion is attributed to a reduction in the mean Y-Mo non-bonded distances and Y-O-Mo bond-angles with increasing temperature, and to their combined effect upon high-energy optical modes and low-energy translational and librational modes. The lowest optical branch corresponding to the translational mode of the oxygen bridging atom in Y-O-Mo linkages has the largest negative Grüneisen parameter and therefore contributes most to the negative thermal expansion behavior[100]. The various vibrational eigenvectors of oxygen atoms relative to yttrium and molybdenum atoms can cause the internal polyhedral to distort unevenly. An anomalously high heat capacity and low thermal conductivity were related to the negative thermal expansion[101]. Water molecules enter freely into $Y_2Mo_3O_{12}$ microchannels and seem to play a role in the partial amorphization of this compound at room temperature. *Ab initio* calculations have been made of water absorption in $Y_2Mo_3O_{12}$, and the absorption geometry of H_2O together with the binding between H_2O and $Y_2Mo_3O_{12}$, have been determined. The water is chemisorbed, with the oxygen atom of the water binding to the Y^{3+} cation. This is further strengthened by hydrogen bonding between each of the hydrogen atoms of the H_2O and the bridging oxygen atom shared by YO_6 and MoO_4 polyhedra. The absorption of water leads to a reduced Y-O-Mo angle and a shortened Y-Mo distance. Volume contraction therefore occurs almost linearly with increasing numbers of water molecules per unit cell, up to eight[102]. Phonon calculations show that the transverse vibration of Y-O-Mo is restricted by water absorption. This in turn hinders negative thermal expansion, since that originates mainly from the vibrational mode. Composites having a tunable coefficient of thermal expansion have been prepared by incorporating orthorhombic yttrium molybdate particles into carbonyl iron particle and silicone resin composites. The porous particles of $Y_2Mo_3O_{12}$ were prepared by using a citric acid method and had a linear coefficient of thermal expansion of -10.4 x 10^{-6}/K between 100 and 800C[103]. The coefficient of thermal expansion of the composite was reduced by about 45% by adding 20wt% of $Y_2Mo_3O_{12}$.

In fully-dense $Al/Y_2Mo_3O_{12}$ composites which had been prepared by squeeze-casting (750C, 1200s, 50MPa), zero thermal expansion was obtained within the temperature range where the thermal mismatch strain was zero. The thermal expansion mismatch of aluminium and $Y_2Mo_3O_{12}$ results in compressive and tensile strains which distort the molybdate lattice[104]. The latter has a positive coefficient of thermal expansion when the surface strain is greater than 0.008%.

$Yb_2(MoO_4)_3$

The compound, $Yb_2Mo_3O_{12}$, was prepared by solid-state reaction, giving an orthorhombic structure which exhibited negative thermal expansion between 200 and 800C. The a- and c-axes exhibited stronger contraction between 200 and 800C, while the b-axis slightly expanded between 200 and 300C and then contracted between 300 and 800C[105]. The linear thermal expansion coefficient was -5.17 x 10^{-6}/K. When powder samples of $Yb_2Mo_3O_{12}$ were prepared by liquid-phase co-precipitation, followed by heat treatment (750C, 6h), the orthorhombic Pbcn structured material had a linear thermal expansion coefficient of -6.237 x 10^{-6}/K[106]. The individual coefficients for the a-axis, b-axis, c-axis and volume were -1.424 x 10^{-6}, -7.599 x 10^{-6}, -9.757 x 10^{-6} and -18.71 x 10^{-6}, respectively.

$ZrMg(MoO_4)_3$

This material crystallizes with orthorhombic symmetry and space group Pnma or $Pna2_1$. It exhibits negative thermal expansion between 300 and 1000K, with a linear coefficient of -3.8 x 10^{-6}/K according to X-ray diffraction data. Between 295 and 775K, the linear coefficient is -3.73 x 10^{-6}/K according to dilatometry[107]. The $ZrMgMo_3O_{12}$ retains an orthorhombic structure, with no phase transition or decomposition, at least from 123K to 1200K. It is not hygroscopic. Looking at the $Sc_2(WO_4)_3$-type Pbcn phases of $Y_2(MoO_4)_3$, $Er_2(MoO_4)_3$ and $Lu_2(MoO_4)_3$ as a whole, polyhedral distortions, transverse vibrations of A $\cdots$ O-Mo (A = Y = rare earth) binding oxygen atoms, non-bonded distances A $\cdots$ O and atomic displacements have been evaluated as a function of the temperature and the ionic radii[108].

$ZrMn(MoO_4)_3$

It was shown that $ZrMnMo_3O_{12}$ has a monoclinic structure, $P2_1/a$, between 298 and 358K and transforms to the orthorhombic structure, Pnma, above 363K. The linear coefficient of thermal expansion, as deduced from X-ray diffraction results, was -2.80 x 10^{-6}/K between 363 and 873K[109]. The coefficient of thermal expansion of a bulk cylinder of the ceramic, as measured using a dilatometer, was -4.7 x 10^{-6}/K between 373 and 773K.

$Zr(MoO_4)_2$

Film samples with a cubic β-ZrW_2O_8 structure have been prepared on glass substrates by means of sol-gel dip-coating. Crystalline $ZrMo_2O_7(OH)_2(H_2O)_2$ precursor powders change into amorphous gel via the coordination reaction of citric acid and polyvinylpyrrolidone with metal ions. A cubic β-$ZrMo_2O_8$ film, consisting of ball-like particles, is obtained upon annealing the gel film, and the bonding strength decreases from 27.8N for gel film to 8.7N for β-$ZrMo_2O_8$ film. The room-temperature lattice parameters are a = b = c = 8.976Å[110]. The thermal expansion coefficient is -8.66 x 10^{-6}/K between room temperature and 430C. The cubic material can also be prepared by carefully controlled dehydration of $ZrMo_2O_7(OH)_2\bullet 2H_2O$, but the quality of the product depends upon the route used to prepare the $ZrMo_2O_7(OH)_2\bullet 2H_2O$; not all acids provide suitable media its preparation. Cubic $ZrMo_2O_8$ can be obtained only from precipitates which contain the hydrate. Starting solutions with a Zr:Mo ratio greater than 1:2 are necessary in order to avoid the co-precipitation of amorphous MoO_3. Some routes lead to material that is contaminated with amorphous impurities[111]. Highly crystalline pure material can be obtained by using a perchloric acid medium for the synthesis of the precursor. Cubic zirconium molybdate has been synthesized by using zirconium oxynitrate and ammonium molybdate as raw materials. Preparation of the metastable phase of cubic $ZrMo_2O_8$ was very sensitive to the heat treatment[112]. Pure cubic $ZrMo_2O_8$ with a linear thermal expansion coefficient of -4.337 x 10^{-6}/K between ambient and about 400C could be prepared.

The heat capacity was measured between 0.6 and 400K. At 298.15K, the standard molar heat capacity was 210.06J/Kmol. The standard molar entropy was 254.3J/Kmol. There was some uncertainty in the value of the entropy, due to the presence of small amounts of chemical and phase impurities[113]. When the heat capacity was compared to the weighted sum of those of the constituent oxides, it was much greater over the entire temperature region.

Cubic γ-$ZrMo_2O_8$ has an average expansion coefficient of -6.9 x 10^{-6}/K between 2 and 200K and of -5.0 x 10^{-6}/K between 250 and 502K[114]. The thermal expansion data provide strong evidence for the presence of dynamic oxygen disorder at above 200K.

Disordered cubic $ZrMo_2O_8$ (Pa3, Z = 4) was studied using high-pressure inelastic neutron scattering at up to 2.5kbar. An observed phonon softening of 0.1 to 0.3meV below 8meV could account for the negative thermal expansion below 100K[115]. The phonon spectrum for the entire energy range up to 150meV was also measured.

High-pressure X-ray diffraction experiments showed that the material undergoes a first-order phase transition, involving an 11% volume decrease, between 0.7 and 2.0GPa, under quasi-hydrostatic conditions[116]. The transition is reversed upon de-compression, but involves considerable hysteresis. Under non-hydrostatic conditions, the material begins to amorphize at above 0.3GPa.

Cubic $ZrMo_{2-x}V_xO_{8-x/2}$ ($0 \leq x \leq 0.5$) compounds having a well-defined nano-rod morphology were synthesized by using a sol-gel process without any surfactant or template. Doping with vanadium did not affect the crystal structure or the negative thermal expansion, but the rod size increased and a narrowed band-gap was observed[117]. All of the compounds exhibited excellent negative thermal expansion properties between 25 and 450C.

Chapter 3

Phosphates

HfMnMo$_2$PO$_{12}$

This new type of negative thermal expansion material, HfMnMo$_2$PO$_{12-\delta}$, prepared by solid-state reaction, exhibits negative thermal expansion between room temperature and 420K. The coefficient of thermal expansion is -3.59 x 10^{-6}/K. The negative expansion is attributed to the coupled rotation of RUMs, caused by transverse vibrations of bridging oxygen atoms and a distortion of polyhedra due to the Jahn-Teller effect and oxygen vacancies. The electrochemical impedance also has a positive thermal coefficient between room temperature and 420K at extremely low frequencies[118]. The coincidence in the temperature range of the two phenomena hints at there existing an underlying relationship; perhaps involving vacancies and/or the Jahn-Teller effect.

NbPO$_5$

First-principles calculations, based upon density functional theory and the quasi-harmonic approximation, of the cell volume of tetrahedral NbPO$_5$ show that it exhibits negative thermal expansion along the a-axis between 473 and 800K. The corresponding average coefficient of thermal expansion was about -0.766 x 10^{-6}/K. The c-axis parameter and cell volume exhibited positive thermal expansion[119]. Vibrational mode analysis, together with Grüneisen-parameter calculations, showed that transverse vibration of oxygen corner atoms accompanied the rocking motions of corner-shared NbO$_6$ octahedra, and that PO$_4$ tetrahedra governed the negative thermal properties of tetrahedral material.

A study of tetragonal NbOPO$_4$ between room temperature and 500C reveals a phase transition at 292C; the space group being P4/n below the transition and P4/nmm above it. The c-axis exhibits positive thermal expansion over the entire temperature range. The a- and b-axes exhibit positive expansion below the transition, and negative expansion above it[120]. The negative thermal expansion is attributed to rocking motions of the corner-sharing NbO$_6$ octahedra and PO$_4$ tetrahedra[121]. The connectivity of the octahedra and tetrahedra remains the same through the transition but, above the transition, the symmetry is orthorhombic and Z decreases by a factor of 2. The negative thermal expansion behavior can also be viewed as being related to the transverse thermal motion of oxygen atoms.

NbTi(PO$_4$)$_3$

This material exhibits considerable volume contraction between 20 and 700C. The coefficients of thermal expansion are between -4.34 and -0.11 x 10^{-6}/K and between 0.08 and 3.07 x 10^{-6}/K for the a- and c-axes, respectively. The similar material, NaTi$_2$(PO$_4$)$_3$, exhibits only positive volume expansion[122]. The difference in behaviour is attributed to the presence of filled sites, in one material, which are empty in the other.

Th$_2$O(PO$_4$)$_2$

High-temperature X-ray diffraction studies of this orthorhombic material revealed a continuous linear thermal contraction of -1.6 x 10^{-6}/K between 20 and 600C, and a near-zero expansion at higher temperatures which resulted from structural deformation involving oxygen-atom oscillation and inter-cationic repulsion[123]. Similar mechanisms are observed in the analogous compounds, Zr$_2$O(PO$_4$)$_2$ (coefficient of 1.5 x 10^{-6}/K) and U$_2$O(PO$_4$)$_2$ (coefficient of -1.4 x 10^{-6}/K); but those found for Th$_2$O(PO$_4$)$_2$ are particularly striking because of the large ionic radius of tetravalent thorium.

Compounds of the forms: A$_{5-4x}$Zr$_x$Zr(PO$_4$)$_3$ (A = Na, K; $0 \leq x \leq 1$), A$_{5-3x}$R$_x$Zr(PO$_4$)$_3$ (R = Al, Fe, Dy, Ho; $0 \leq x \leq 1.33$) and A$_{5-2x}$B$_x$Zr(PO$_4$)$_3$ (B = Mg, Ca, Sr; $0 \leq x \leq 2$), have been studied between 293 and 1023K. They are members of a large family of structural analogues of NaZr$_2$(PO$_4$)$_3$ and have the general formula, (M$_1$)(M$_2$)$_3$L$_2$(PO$_4$)$_3$, where the L and M atoms are located within the framework and framework voids, respectively[124]. The present phosphates have rhombohedral symmetry; with some being R$\bar{3}$c and others R$\bar{3}$2. The above compounds exhibited thermal expansion anisotropy: with negative expansion occurring along the a-axis and positive expansion occurring along the c-axis. Some of the K-Zr-phosphates exhibit a net negative or near-zero thermal expansion. In order to determine the origin of negative thermal expansion in materials such as ZrV$_x$P$_{2-x}$O$_7$ x = 0 to 2) and AW$_2$O$_8$ (A = Zr, Hf), *ab initio* self-consistent field molecular orbital calculations were performed for the model compound, (HO)$_3$P-O-P(OH)$_3{}^{2+}$. The results suggested that symmetrical stretching vibrations of doubly coordinate oxygen bridges are strongly coupled to their bending vibrations, and are more important to negative thermal expansion at high temperatures. The oxygen bridges accommodate large amounts of stretching of the A-O (A = Zr, Hf) and M-O (M = P, V) bonds at high temperatures by decreasing the bond angles of the oxygen bridges[125]. When the bond angles increase, the bonds are effectively shortened due to the π-type orbital interactions that are associated with the oxygen lone pair orbitals.

$Zr_2(MoO_4)(PO_4)_2$

This material is orthorhombic, with a $Sc_2W_3O_{12}$ structure, between 9 and 400K. It exhibits anisotropic thermal expansion, with coefficients of -8.35 x 10^{-6}, 3.25 x 10^{-6} and -8.27 x 10^{-6}/K for the a-axis, b-axis and c-axis, respectively, between 122 and 400K. These values are similar in magnitude to those of $A_2M_3O_{12}$ (M = Mo or W) compounds with large A^{3+} additions[126]. The contraction on heating is attributed to Zr-O-Mo/P bond-angle changes, similar to those in $Sc_2W_3O_{12}$. During heating, the most marked reductions in separation between the positions of neighboring zirconium and phosphorus atoms are not associated with significant reductions in the corresponding Zr-O-P crystallographic bond angles, contrary to the case of $Sc_2W_3O_{12}$.

$ZrScW_2PO_{12}$

This novel material exhibits negative thermal expansion between 138 and 1300K, and an intense photoluminescence. Temperature-dependent Raman and photoluminescence spectra indicate that it has an orthorhombic structure down to about 74K. The intense photoluminescence can be divided into four bands, reflecting different shifts with increasing temperature[127]. The optical properties are attributed to an n- and p-type co-doping effect, and the detailed behavior ascribed to the abnormal thermal expansion.

$Zr_2(WO_4)(PO_4)_2$

The material can be prepared from ZrO_2, WO_3 and P_2O_5 by one-step sintering for 1h, and the relative density can attain 99.8%. The grain size is smaller and more uniform when MgO is added to the raw materials than when MgO is added to $Zr_2(WO_4)(PO_4)_2$ powder[128]. The coefficient of thermal expansion was -2.325 x 10^{-6}/K for samples without MgO, -1.406 x 10^{-6}/K when MgO was added to the raw materials, -1.509 x 10^{-6}/K when MgO was added to $Zr_2(WO_4)(PO_4)_2$ powder and -1.384 x 10^{-6}/K when MgO and PVA were added to the raw materials. Magnesium composites containing $Zr_2(WO_4)(PO_4)_2$ particles were fabricated by extruding the component powders. The coefficient of thermal expansion of the composites decreased with increasing fraction of ceramic, but the effect was insignificant; perhaps because of the low bulk modulus of the ceramic. There existed a tension-compression asymmetry in the room-temperature yield strength of magnesium without the phosphate[129]. This was reduced in the composites, and the change was attributed to the introduction of residual tensile stresses into the matrix during fabrication, due to the negative thermal expansion of the phosphate. Negative thermal expansion materials can be subjected to high stresses when incorporated into composites. Under ambient conditions, this material exhibits anisotropic negative thermal expansion, with values of: -14.0 x 10^{-6}, -7.9 x 10^{-6}, 2.5 x 10^{-6} and -8.7 x 10^{-6}/K for the volume, a-axis, b-axis and c-axis, respectively, at 0GPa. The effect of hydrostatic pressure was investigated between 300 and 60K under pressures of up to 0.3GPa. No phase transitions were observed[130]. The material had a bulk modulus of 61.3GPa at room temperature, and pressure had no detectable effect upon the thermal expansion; with values of -14.2 x 10^{-6}, -7.9 x 10^{-6}, 2.9 x 10^{-6} and -9.2 x 10^{-6}/K for the volume, a-axis, b-axis and c-axis, respectively, at 0.3GPa.

$Zr_2W_2P_2O_{15}$

Samples of $Zr_2W_2P_2O_{15}$ having a thermal expansion coefficient of -4.01 x 10^{-6}/K between 143 and 673K were prepared by controlling the reaction so as to avoid the intermediate phase, $Zr_2WP_2O_{12}$, whose inferior negative coefficient of expansion limits its application. Between 1473 and 1573K, $Zr_2WP_2O_{12}$ forms easily but the reaction between $Zr_2WP_2O_{12}$ and WO_3 to generate $Zr_2W_2P_2O_{15}$ is difficult, even when followed by heating at up to 1673K[131]. By directly putting ZrO_2, WO_3 and $NH_4H_2PO_4$ into a pipe furnace at the sintering temperature of 1673K, $Zr_2W_2P_2O_{15}$ is formed while avoiding the intermediate phase, $Zr_2WP_2O_{12}$. Powder was synthesized by using the solid-state reaction method and optimum heating conditions (1200C, 4h). The relative density of $Zr_2WP_2O_{12}$ ceramics with no sintering additive was 60%. That of samples sintered together with more than 0.5mass%MgO was about 97%. The average grain size of a sample sintered at 1200C for 4h was about 1μm[132]. The resultant ceramics had a thermal expansion coefficient of about -3.4 x 10^{-6}/K. The Young's modulus, Poisson ratio, three-point bending strength, Vickers microhardness and fracture toughness were 74GPa, 0.25, 113MPa, 4.4GPa and 2.3MPam$^{1/2}$, respectively.

Powder prepared by hydrothermal means was used as a filler to tailor the thermal expansion of polyimide-based composites. A series of composites, containing various amounts (0 to 40wt% or 0 to 19.6vol%) of the powder were fabricated via *in situ* polymerization[133]. Addition of the powder steadily reduced the thermal expansion of the matrix at all loadings. 40wt% of the powder led to a 32.5% (15 x 10^{-6}/K) reduction in the expansion coefficient. The thermal stability of the composites increased with increasing powder content.

Chapter 4

Vanadates

$Cu_2V_2O_7$

The orthorhombic α-$Cu_2V_2O_7$, prepared using a solid-state method, exhibited negative thermal expansion between room temperature and 470K with a coefficient of thermal expansion of -5.8 x 10^{-6}/K. The negative expansion behaviour was attributed to the coupled rotation of rigid unit modes caused by transverse vibrations of bridge oxygen atoms. The negative expansion was non-linear between 470 and 610K, where the average coefficient was about -2.4 x 10^{-6}/K. This was caused by the Jahn-Teller effect and a small amount of microporosity[134]. The material exhibited low thermal expansion between 610 and 720K, with a coefficient of 0.43 x 10^{-6}/K.

$HfScMo_2VO_{12}$

Structural analysis shows that this material has the orthorhombic structure, Pbcn. The linear coefficient of thermal expansion is -2.78 x 10^{-6}/K between -150 and 675K and -2.99 x 10^{-6}/K between 300 and 780K, according to dilatometry, and -2.11 x 10^{-6}/K according to temperature-dependent X-ray diffraction[135]. The material also exhibits intense wide-band photoluminescence over the visible region. Other studies indicate an intrinsic coefficient of thermal expansion of -1.27 x 10^{-6}/K between 140 and 1469K[136].

$NbVO_5$

Nanocrystalline $NbVO_5$ was prepared by co-precipitating Nb_2O_5 and NH_4VO_3 and calcining (550C, 2h). The compound is completely formed at 504.5C, and thermally stable below 658C. Its structure is orthorhombic, Pnma, with a = 11.8453, b = 5.5126 and c = 6.9212Å. Negative thermal expansion with a coefficient of -6.63 x 10^{-6}/K occurs between room temperature and 600C[137]. This is attributed to tilting of the NbO_6 octahedra and VO_4 tetrahedra within the flexible framework structure.

TaVO$_5$

This framework structure exhibits negative thermal expansion at up to 1073K, with a reversible low-temperature transition to a monoclinic phase at 259K. Under a pressure of 0.2GPa, it changes to a phase which is probably the same as the low-temperature phase; suggesting that this structural phase transition may be volume-driven[138]. X-ray diffraction and neutron powder diffraction data confirm a 259K transformation from monoclinic symmetry, P2$_1$/c, to orthorhombic, Pnma. The rigid TaO$_6$ octahedra in the orthorhombic phase become non-regular at 295K, resulting in the above transition. The material exhibits negative thermal expansion above room temperature: the volumetric thermal expansion coefficient is -8.92 x 10^{-6}/K between 20 and 600C, and -21.9 x 10^{-6}/K above 600C[139]. The negative thermal expansion behavior is attributed to the tilt of the TaO$_6$ and VO$_4$ polyhedra, where shrinkage of the VO$_4$ tetrahedra results in an increase in Ta-O-V angles during heating, while the Ta-O$_1$-Ta angle remains at 180° in the framework.

ZrScMo$_2$VO$_{12}$

The novel material ZrScMo$_2$VO$_{12}$, with an orthorhombic Pbcn structure at room temperature, exhibits high negative thermal expansion over the wide range of 150 to 823K. A transition from monoclinic to orthorhombic occurs between 70 and 90K[140]. Its intense white-light photoemission is attributed to an n- and p-type co-doping effect which creates both donor- and acceptor-like states in the band gap, as well as donor-acceptor pairs and even bound exciton complexes.

ZrV$_2$O$_7$

Both ZrV$_2$O$_7$ and HfV$_2$O$_7$ exhibit negative thermal expansion in the high-temperature phases. Two marked anomalies, due to successive phase transitions, are observed in the temperature dependences of the heat capacity: at 345.5 and 373.4K for ZrV$_2$O$_7$ and at 341.8 and 370.3K for HfV$_2$O$_7$. The effective phonon densities of states were described by a simple model, and the mode-Grüneisen parameters of the low-temperature phase were deduced from the heat capacities. The effective phonon densities of state typified the common features of negative expansion compounds: low-energy phonon mode, high-energy phonon mode and wide intermediate gap. The mode-Grüneisen parameter of low-energy modes corresponding to translational and librational vibrations of the constituent polyhedra is negative, but with a small absolute value due to the distortion of V$_2$O$_7$ groups in the low-temperature phase. This results in a positive thermal expansion[141]. The release of structural distortion, during successive phase transitions with a large volume increase, leads to the negative thermal expansion of ZrV$_2$O$_7$ and HfV$_2$O$_7$ in the high-temperature phase. Solid solutions of the form, Zr$_{1-x}$Hf$_x$V$_2$O$_7$ (x = 0 to 1), were all monophase and had a cubic structure. The room-temperature lattice parameters decreased linearly with increasing hafnium content due to the smaller ionic radius of Hf^{IV} as compared with that of Zr^{IV}. Sudden positive thermal expansions occurred in all of the samples near to 350 and 375K. Negative thermal expansion occurred above about 380K. All of the samples also underwent two successive phase transitions between 330 and 390K[142]. The transition temperatures decreased with increasing x-value content; thus reflecting a decrease in superlattice structure.

Microfibers of ZrV_2O_7 with diameters of 1 to 3μm were prepared by using a sol-gel technique, and compared with ZrV_2O_7 powders which had been prepared by using the same method[143]. The pH value had little effect upon the crystal structure of the product, but had a significant effect upon the morphology. Fibers which were obtained using a pH of 9 had a cylindrical morphology and a diameter of about 1μm. Fibers having such a cylindrical morphology, and irregular powders with an average size of between 100 and 200nm, both exhibited negative thermal expansion between 150 and 600C[144]. The as-prepared fibers first exhibited positive thermal expansion, and then a negative thermal expansion which resulted from a phase transition from a 3 x 3 x 3 superstructure to a 1 x 1 x 1 cubic structure.

Lattice dynamics calculations have been made of the negative thermal expansion behavior of the ZrV_2O_7 family. The calculations quantitatively reproduced the negative expansion over a range of temperatures[145]. The calculations also revealed a soft-phonon mode at a wave vector of 0.31<110>, which was in excellent agreement with the known incommensurate modulation in ZrV_2O_7 below 375K.

Compounds chosen from the ZrO_2-V_2O_5-MoO_3 phase diagram were prepared by using the solution combustion method. Single crystals of $ZrV_{1.50}Mo_{0.50}O_{7.25}$, were grown by melt-cooling. They belonged to the cubic crystal system, $Pa\overline{3}$, with a = 8.8969Å, V = 704.24Å^3 and Z = 4. The Zr^{4+} occupies the 4a position, while V^{5+} and Mo^{6+} occupy two 8c Wyckoff positions. Two fully-occupied oxygen atoms, (24d and 4b) and one partially occupied oxygen atom (8c) were identified[146]. The structure is related to those of ZrV_2O_7 and cubic $ZrMo_2O_8$. The single crystals exhibited negative thermal expansion above 370K. Zirconium vanadates having a chemical composition which included niobium and yttrium were prepared by using the sol-gel method[147]. Although ZrV_2O_7 exhibits a generally positive thermal expansion at above room temperature, the degree of positive thermal expansion decreased upon substituting phosphorus for vanadium and niobium plus yttrium for zirconium.

Phases in vanadates of the form, $ZrP_{2-x}V_xO_7$, exhibit isotropic thermal expansion which can be negative, over wide temperature ranges, up to at least 950C. A phase transition occurs in ZrP_2O_7 at 290C: the thermal expansion is normal below this temperature, but is very small and positive above it. Two phase transitions occur in ZrV_2O_7, close to 100C. Each of them results in a sudden increase in volume with increasing temperature. Markedly negative thermal expansion begins in ZrV_2O_7 at above 100C and continues up to about 800C. Decomposition then occurs. Below the phase transitions in cubic ZrP_2O_7 and ZrV_2O_7, a 3x3x3 superstructure appears, with no apparent change in space group. In the high-temperature structures, the V-O-V and P-O-P angles are constrained, by the Pa3 space group symmetry, to be equal to 180°. When the 3x3x3 superstructure has developed, the above symmetry constraint is relaxed for some of the V-O-V and P-O-P linkages. The 3x3x3 superstructure results in eleven crystallographic sites for vanadium or phosphorus. The thermal expansion is related to the frustration involved in bending the V-O-V or P-O-P angles away from 180° in a cooperative way[148]. Among the middle members of the ZrV_2O_7-ZrP_2O_7 solid solutions, phase transitions above room temperature are entirely suppressed. In $ZrP_{2-x}V_xO_7$ compounds, there is a preponderance of VPO_7^{4-} groups, relative to a mixture of V_2O_7 and P_2O_7 groups, when x is close to unity.

Structural analysis shows that $Zr_{0.70}V_{1.33}Mo_{0.67}O_{6.73}$ has the cubic crystal structure, Pa3, at room temperature. It exhibits stable isotropic negative thermal expansion, and maintains a cubic structure from 103 to 773K; with no phase transition[149]. The average linear coefficients of thermal expansion are -3.75×10^{-6}/K, between 163 and 673K, and -4.50×10^{-6}/K, between room temperature and 773K. The phase transition temperature of ZrV_2O_7 is reduced to below room temperature by adding $ZrMo_2O_8$. The negative expansion of $Zr_{0.70}V_{1.33}Mo_{0.67}O_{6.73}$ is attributed to the quasi-rigid unit modes. Tungsten-doped ZrV_2O_7 has been synthesized using the sol-gel method. The presence of W^{6+} causes the stabilization of phases having a 1x1x1 cubic structure, Pa$\bar{3}$ (Z = 4). Specimens of $ZrV_{1.6}W_{0.4}O_{7.2}$ exhibit negative thermal expansion at room temperature whereas $ZrV_{2-x}W_xO_{7+\delta}$ generally exhibits positive thermal expansion[150].

25

Chapter 5

Tungstates

This is perhaps the most important group of negative thermal expansion materials, as they have potential application in the fields of electronics, aeronautics and optics[151]. Typical examples of this class are $Sc_2W_3O_{12}$, $Lu_2W_3O_{12}$ and $Y_2W_3O_{12}$ where the most commonly cited negative thermal expansion mechanism is the so-called quasi-rigid unit model.

$Al_2(WO_4)_3$

Neutron powder diffraction studies, performed between 20 and 800C, showed that the coefficients of thermal expansion were -1.31×10^{-6}, 5.94×10^{-6} and $-9.94 \times 10^{-6}/K$ for the a-axis, b-axis and c-axis, respectively[152]. When films were grown by sequential magnetron sputtering of WO_3 and Al_2O_3 targets onto quartz substrates, the annealing temperature affected the microstructure and stoichiometry of the film. Smooth compact films were amorphous, with quite strong interfacial adhesion. With increasing annealing temperature, grains grew with rather a high surface defect density. During heating from room temperature to 800C, the unit-cell volume first increased and then decreased. The films exhibited an anisotropic negative thermal expansion[153]: the average thermal expansion coefficients for a-, b- and c-axes being -3.25×10^{-6}, 5.15×10^{-6} and $-2.64 \times 10^{-6}/K$, respectively, while the average volume thermal expansion was $-0.69 \times 10^{-6}/K$. Lattice dynamics calculations have been made of the orthorhombic phases of $Al_2(WO_4)_3$ and $Sc_2(MoO_4)_3$ by using a rigid-ion model to calculate the phonon frequencies at the Brillouin zone center. The results were a qualitative fit to the Raman spectra observed in these materials at ambient temperature and pressure. A higher coupling was observed between the modes of $Al_2(WO_4)_3$ as compared with $Sc_2(MoO_4)_3$. Pressure-dependent studies showed that the 362 and 433/cm modes of $Sc_2(MoO_4)_3$ and $Al_2(WO_4)_3$, respectively, had unusually high $\partial\omega/\partial P$ values[154]. The presence of these unstable modes was tentatively related to the phase transition from the monoclinic high-pressure phase to a phase of lower symmetry.

$Cr_2(WO_4)_3$

Powder samples of $Cr_2(WO_4)_3$ were prepared by liquid-phase reaction sintering. X-ray diffraction data for 298 to 1073K revealed an essentially linear increase in cell volume as a function of temperature. The intrinsic linear coefficient of thermal expansion which was deduced from these data was 1.274×10^{-6}/K. When the thermal expansion behavior was studied in static air using a thermo-dilatometer there was an initial positive thermal expansion, followed by a phase transition and then a negative thermal expansion[155]. The thermal expansion coefficient here was -7.033×10^{-6}/K.

$Dy_2(WO_4)_3$

The phase transition of $Dy_2W_3O_{12}$ from monoclinic to orthorhombic was shown to occur at 996C. The orthorhombic phase could be retained by quenching. The coefficient of thermal expansion was -26×10^{-6}/K at 150 to 500C; the largest negative thermal expansion in the $A_2W_3O_{12}$ family, where A is a rare-earth element[156]. A proposed negative expansion mechanism involved transverse thermal motion of the bridge oxygen in A-O-W linkages, together with distortion of the polyhedra with large ions such as Dy^{3+} on the A-site.

$Er_2(WO_4)_3$

Negative thermal expansion in the orthorhombic tungstate was measured by high-temperature X-ray diffraction[157]. The coefficients of thermal expansion were -10.14×10^{-6}, -3.35×10^{-6}, -6.70×10^{-6} and -20.22×10^{-6}/K for the a-axis, b-axis, c-axis and volume, respectively, between 473 and 1073K (table 2). When solid solutions of the form, $Er_2W_{3-x}Mo_xO_{12}$ ($0.5 \leq x \leq 2.5$), were synthesized using solid-state methods, all of the samples had the same orthorhombic structure and exhibited a negative thermal expansion which was related to the transverse vibration of bridging oxygen atoms. The thermal expansion coefficients of $Er_2W_{3-x}Mo_xO_{12}$ were -16.2×10^{-6}/K for x = 0.5 and -16.5×10^{-6}/K for x = 2.5 between 200 and 800C. They were -20.2×10^{-6} and -18.4×10^{-6}/K for the end-members, $Er_2W_3O_{12}$ and $Er_2Mo_3O_{12}$ respectively, over the same temperature range[158]. High-temperature X-ray diffraction data suggested that the difference between W-O and Mo-O was responsible for the change in thermal expansion due to element substitution.

$Hf(WO_4)_2$

Densely-packed hafnium tungstate blocks have been synthesized by rapid solidification using a CO_2 laser, giving a cubic structure with minor amounts of orthorhombic phase. Specific Raman bands appeared in the samples synthesized using lasers but not in samples prepared using solid-state reaction[159]. This was attributed to a compressive stress which was induced during rapid solidification, due to a sudden drop in temperature from the molten pool to ambient. Thin films have been grown on quartz substrates by using pulsed-laser deposition between room temperature and 500C under oxygen pressures of 5 to 20Pa. The as-deposited films comprise amorphous phases. Crystallized cubic HfW_2O_8 films can be prepared by heating (1200C, 420s). Thin films which were grown at 500C under an oxygen pressure of 5Pa had the smoothest and most uniform surfaces. An α-to-β phase transition occurs between 100 and 150C[160]. The average linear thermal expansion coefficient is -9.33 x 10^{-6}/K between 25 and 600C. Cubic HfW_2O_8 has also been prepared by the solid-state reaction of HfO_2 and WO_3. The ceramic can be prepared via sintering (1200C, 6h) and quenching in de-ionized water. The product is compact, with small square grains. An α-to-β transition occurs at 182.5C, leading to a decrease in the thermal expansion coefficient[161]. The coefficient is -12.90 x 10^{-6}/K for α-HfW_2O_8, and -10.09 x 10^{-6}/K for β-HfW_2O_8. The average coefficient is -11.46 x 10^{-6}/K between 25 and 600C.

The heat capacities of ZrW_2O_8 and HfW_2O_8 have been measured between 1.8 and 330K, showing that the curves for both materials are very similar. The heat capacity of HfW_2O_8 at low temperatures is larger than that of ZrW_2O_8 due to atomic mass effects, but the heat capacities are equal at about 220K. The frequency distributions of the lattice vibrations were estimated by analysing the heat capacities[162]. A difference in the frequency distributions arose from the differing atomic masses and bond strengths.

The negative thermal expansion of HfW_2O_8 is related to transverse thermal vibrations of the bridging oxygen atoms[163]. These lead to coupled rotations of the essentially rigid polyhedral building blocks of the structure. The material undergoes marked softening, in which the average bulk modulus falls from 69GPa at 298K to 48GPa at 430K, as the WO_4 orientation order-disorder transition temperature approaches. This is associated with an increasingly negative thermal expansion, and reversible WO_4 orientational disordering under compression. The α-HfW_2O_8 becomes elastically softer upon compression at constant temperature, and the $\alpha \rightarrow \beta$ phase transition temperature decreases by about 30K between 52 and 414MPa[164]. Above this phase transition, no further temperature-dependent softening or pressure-dependent changes occur in the coefficient of thermal expansion.

A marked softening of low-energy phonons in ZrW_2O_8 is responsible for its anomalous thermal expansion behavior. In order to understand the effect of replacing zirconium with hafnium, upon the negative thermal expansion behavior, lattice dynamic calculations and neutron time-of-flight spectroscopic measurements were made of the phonon density of states for cubic HfW_2O_8. The calculated phonon spectrum was in reasonable agreement with the experimental data[165]. The phonon spectra in the zirconium and hafnium compounds differed at low energies, largely due to the mass difference. The heat capacity of HfW_2O_8 was measured between 1.8 and 330K and the Grüneisen function was deduced for 80 to 330K from the heat capacity and other thermal data. The effective phonon density-of-states was found from an analysis of the heat capacity; assuming one Debye, two Einstein and two rectangular-type functions[166]. The mode Grüneisen parameters were estimated from the phonon distribution, showing that two low-energy Einstein modes at 3 to 6meV have large negative mode Grüneisen parameters; resulting in the observed negative thermal expansion.

Table 2 Thermal Expansion Coefficients of Rare-Earth Tungstates.

Composition	Temperature Range (K)	Axis	Coefficient (/K)
$Y_2W_3O_{12}$	473-1073	a	-9.78×10^{-6}
$Y_2W_3O_{12}$	473-1073	b	-5.13×10^{-6}
$Y_2W_3O_{12}$	473-1073	c	-6.68×10^{-6}
$Y_2W_3O_{12}$	473-1173	a	-7.38×10^{-6}
$Y_2W_3O_{12}$	473-1173	b	-3.42×10^{-6}
$Y_2W_3O_{12}$	473-1173	c	-6.20×10^{-6}
$Er_2W_3O_{12}$	473-1073	a	-10.14×10^{-6}
$Er_2W_3O_{12}$	473-1073	b	-3.35×10^{-6}
$Er_2W_3O_{12}$	473-1073	c	-6.70×10^{-6}
$Yb_2W_3O_{12}$	473-1073	a	-10.20×10^{-6}
$Yb_2W_3O_{12}$	473-1073	b	-2.65×10^{-6}
$Yb_2W_3O_{12}$	473-1073	c	-6.41×10^{-6}
$Lu_2W_3O_{12}$	473-1073	a	-9.70×10^{-6}
$Lu_2W_3O_{12}$	473-1073	b	-2.89×10^{-6}
$Lu_2W_3O_{12}$	473-1073	c	-5.74×10^{-6}

$Lu_2W_3O_{12}$	393-893	a	-9.90 x 10^{-6}
$Lu_2W_3O_{12}$	393-893	b	-2.20 x 10^{-6}
$Lu_2W_3O_{12}$	393-893	c	-8.30 x 10^{-6}
$Sc_2W_3O_{12}$	10-450	a	-6.50 x 10^{-6}
$Sc_2W_3O_{12}$	10-450	b	5.63 x 10^{-6}
$Sc_2W_3O_{12}$	10-450	c	-5.74 x 10^{-6}

The effect of pressure upon the crystal structure of HfW_2O_8 was investigated using neutron powder diffraction. Under a hydrostatic pressure of 0.62GPa at room temperature, the cubic material transforms, with a 5% reduction in volume, to the same orthorhombic phase seen in ZrW_2O_8 at above 0.21GPa. The sluggish transformation requires some 24h to complete at constant pressure but, once formed, the orthorhombic phase is retained after removing the pressure. Upon heating to 360K, the metastable orthorhombic phase transforms back to cubic[167]. The much higher pressure required for the cubic-to-orthorhombic transition in HfW_2O_8, as compared with ZrW_2O_8, is suggested to be important when the present material is used in composites, because of the large local pressures which can occur.

It is noted that $HfMgMo_3O_{12}$ exhibits low positive thermal expansion while $HfMgW_3O_{12}$ exhibits low negative thermal expansion. The solid solution, $HfMgMo_{1.5}W_{1.5}O_{12}$, exhibits an enhanced negative thermal expansion, with a lower anisotropy with respect to both $HfMgMo_3O_{12}$ and $HfMgW_3O_{12}$. The coefficients of thermal expansion for $HfMgMo_{1.5}W_{1.5}O_{12}$ are -6.78 x 10^{-6}, 6.59 x 10^{-6}, -5.23 x 10^{-6} and -6.19 x 10^{-6}/K for the a-axis, b-axis, c-axis and volume, respectively. This yields a linear coefficient of -2.06 x 10^{-6}/K[168]. The alternative arrangements of tungsten and molybdenum around each octahedra, and differences between them with regard to electronegativity and cation radius, were suggested to explain the enhanced expansion and decreased anisotropy with respect to $HfMgMo_3O_{12}$.

Table 3 Thermal Expansion Coefficients of $HfW_{2-x}V_xO_{8-x/2}$ Compositions.

Composition	Temperature Range (K)	Coefficient (/K)
HfW_2O_8	490-573	-5.1×10^{-6}
$HfW_{1.90}V_{0.10}O_{7.95}$	460-573	-4.5×10^{-6}
$HfW_{1.85}V_{0.15}O_{7.925}$	431-573	-4.5×10^{-6}
$HfW_{1.80}V_{0.10}O_{7.90}$	300-573	-4.2×10^{-6}
$HfW_{1.75}V_{0.25}O_{7.875}$	380-573	-3.7×10^{-6}
$HfW_{1.70}V_{0.30}O_{7.85}$	350-573	-3.4×10^{-6}

Powder X-ray diffraction studies of cubic $Zr_{1-x}Hf_xW_2O_8$ (x = 0.25, 0.50 and 0.75) solid solutions at 90 to 560K showed that the lattice parameters at 121K decreased linearly with increasing hafnium content, due to its smaller ionic radius compared to that of zirconium[169]. The temperature of the α-β phase transition increased with increasing hafnium contents, reflecting the decrease in lattice free volume related to the orientation of the unshared vertex of WO_4. The transition entropy of $Zr_{0.5}Hf_{0.5}W_2O_8$ was 2.1J/molK; consistent with those of ZrW_2O_8 and HfW_2O_8 and suggesting that $Zr_{1-x}Hf_xW_2O_8$ (x = 0 to 1.0) has the same order-disorder phase transition mechanism.

Single-phase orthorhombic $(HfMg)(WO_4)_3$, an $A_2(WO_4)_3$-type tungstate, was prepared by calcining HfO_2, MgO and WO_3; thus substituting Hf^{4+} and Mg^{2+} for the A^{3+} cations in $A_2(WO_4)_3$. The resultant material had a thermal expansion coefficient of about -2×10^{-6}/K between room temperature and 800C[170]. The negative thermal expansion mechanism was assumed to be the same as that in $Sc_2(WO_4)_3$. Magnesium hafnium tungstate was synthesized by means of high-energy ball-milling, again followed by calcination. It crystallized in the space group, $P2_1/c$, below 400K and transformed to an orthorhombic structure at higher temperatures; with the space group being Pnma rather than the usual Pnca one of other $A_2(MO_4)_3$ materials. Negative thermal expansion was observed in the orthorhombic phase, with coefficients of -5.2×10^{-6}, 4.4×10^{-6}, -2.9×10^{-6} and -3.7×10^{-6}/K for the a-axis, b-axis, c-axis and volume, respectively[171]. The monoclinic to orthorhombic phase transition involved a smooth change in the unit-cell volume; indicating a second-order phase transition.

Cubic $HfW_{2-x}V_xO_{8-x/2}$ solid solutions, with x up to 0.34, were synthesized by using acidic steam hydrothermal-thermal dehydrating methods. This produced tetragonal $HfW_{2-x}V_xO_{7-x/2}(OH)_2(H_2O)_2$ after the acidic steam hydrothermal treatment and orthorhombic $HfW_{2-x}V_xO_{8-x/2}$ after dehydration. The cubic phase was obtained at higher temperatures[172]. The crystal structures of the cubic phase were of α-HfW_2O_8 and β-HfW_2O_8 type for $0 \leq x \leq$ 0.30 and x = 0.34, respectively. All of the cubic specimens exhibited high negative thermal expansions (table 3).

$In_2(WO_4)_3$

Samples have been prepared from In_2O_3 and WO_3 by solid-state reaction. Monoclinic $In_2W_3O_{12}$ was obtained by sintering (900C, 6h) and consisted of uniform grains with an average size of 4 to 6μm. A transition from monoclinic to orthorhombic occurred at 253.34C. The monoclinic material exhibited positive thermal expansion, with an average linear thermal expansion coefficient of 16.51 x 10^{-6}/K between 27 and 249C[173]. The orthorhombic material exhibited negative thermal expansion, with an average linear thermal expansion coefficient of -3.00 x 10^{-6}/K between 273 and 700C. Solid solutions of the form, $In_{2-x}Sc_xW_3O_{12}$ ($0 \leq x \leq 2$), were synthesized using solid-state reaction methods. The $In_2W_3O_{12}$ ceramic underwent a transition from monoclinic to orthorhombic at 248C. This phase transition temperature was easily moved to a lower temperature by partially replacing the In^{3+} with Sc^{3+}. As the x-value was increased from 0 to 1, the phase transition temperatures of $In_{2-x}Sc_xW_3O_{12}$ ($0 \leq x \leq 2$) decreased from 248 to 47C. All of the $In_{2-x}Sc_xW_3O_{12}$ ($0 \leq x \leq 2$) exhibited negative thermal expansion below the corresponding phase transition temperature[174]. The thermal expansion coefficients changed from -1.08 x 10^{-6} to -7.13 x 10^{-6}/K.

$Li_2Ni(WO_4)_2$

Neutron diffraction studies show that the Ni^{2+} spins of S = 1 for NiO_6 octahedra are coupled via corner-sharing non-magnetic double tungstate groups via a super super-exchange route. Magnetic anomalies at about 18 and 13K presage the onset of a commensurate long-range antiferromagnetic ordering[175]. Negative thermal expansion occurs below 13K and is attributed to competing normal thermal contraction and a long-range antiferromagnetic spin ordering through counterbalanced WO_4 and NiO_6 local polyhedral distortion.

$Lu_2(WO_4)_3$

Negative thermal expansion in the orthorhombic tungstate was measured by means of high-temperature X-ray diffraction[176]. The coefficients of thermal expansion were -9.70 x

10^{-6}, -2.89 x 10^{-6}, -5.74 x 10^{-6} and -18.54 x 10^{-6}/K for the a-axis, b-axis, c-axis and volume, respectively, between 473 and 1073K. The negative thermal expansion of the $Sc_2W_3O_{12}$ family has been made much more pronounced by substituting a larger cation for the scandium. X-ray diffraction measurements, performed at 127 to 627C, indicate a linear thermal expansion coefficient of -6.8 x 10^{-6}/K for $Lu_2W_3O_{12}$, as compared to -2.2 x 10^{-6}/K for $Sc_2W_3O_{12}$. Negative thermal expansion in this family of materials depends upon the rocking motions of polyhedra, but the polyhedra in this structure cannot rock without changing shape[177]. The presence of larger cations expands the octahedra, thus reducing oxygen-oxygen repulsion within them and facilitating the polyhedra shape-changes which are necessary to permit the rocking motions required for negative thermal expansion. Solid solutions of the form, $Lu_2W_{3-x}Mo_xO_{12}$ ($0.5 \leq x \leq 2.5$), were prepared by using solid-state methods. They all had an orthorhombic structure and exhibited a negative thermal expansion which was related to the transverse vibration of bridging oxygen atoms in the structure. The thermal expansion coefficients of $Lu_2W_{3-x}Mo_xO_{12}$ were -20.0 x 10^{-6}/K for x = 0.5 and -16.1 x 10^{-6}/K for x = 2.5, as compared with -18.6 x 10^{-6} and -16.9 x 10^{-6}/K for unsubstituted $Lu_2W_3O_{12}$ and $Lu_2Mo_3O_{12}$, respectively, between 200 and 800C[178]. High-temperature X-ray diffraction data indicated that the difference between the W-O and Mo-O bonds was responsible for the change in thermal expansion coefficient following element substitution.

$Sc_2(WO_4)_3$

Negative thermal expansion has been discovered in a very large family of oxides having the formula, $A_2M_3O_{12}$, where the A and M cations are coordinated octahedrally and tetrahedrally, respectively, by oxygen. In $Sc_2W_3O_{12}$, negative thermal expansion occurs over a wide temperature range[179]. The trends which are observed in this large family have provided some insight into why many compounds which exhibit negative thermal expansion cease to do so below a certain temperature.

Powders of $Sc_2(WO_4)_3$ were synthesized by the solid state reaction of scandium dioxide and tungsten trioxide, giving high-purity orthorhombic material. The particles were irregular and were between 0.2 and 0.7μm. They underwent no phase transition or decomposition between room temperature and 1200C[180]. The average microscopic and macroscopic thermal expansion coefficients were -1.9 x 10^{-6} and -5.6 x 10^{-6}/K, respectively, between 30 and 800C. Overall, $Sc_2(WO_4)_3$ exhibits negative thermal expansion between 10 and 1073K. Powder neutron diffraction data for 10 to 450K reveal a linear decrease in cell volume as a function of temperature, with an intrinsic linear coefficient of thermal expansion of -2.2 x 10^{-6}/K. The linear coefficient of thermal expansion, as measured for a ceramic bar of $Sc_2(WO_4)_3$, could be as negative as -11 x

10^{-6}/K; due to microstructural changes as a function of temperature. Rietveld refinement, as a function of temperature, suggests that the intrinsic negative thermal expansion is related to transverse vibrations of the bridging oxygen atoms in the structure. The anharmonic nature of those vibrations leads to coupled tilting of the quasi-rigid framework polyhedra[181]. The tilting then causes the structure to become denser with increasing temperature.

Thin films of $Sc_2W_3O_{12}$ have been grown onto silicon and quartz substrates by means of pulsed laser deposition. The as-deposited films were amorphous. Orthorhombic crystalline films formed upon heating (1000C, 420s, air). Films which were deposited at 500C under an oxygen pressure of 20Pa had the lowest root-mean-square roughness and gave the optimum optical performance, with an average transmission of over 85%[182]. The Young's modulus was 153.40GPa, the hardness was 8.62GPa and anisotropic negative thermal expansion with an average linear expansion coefficient of -3.0 x 10^{-6}/K was found between 25 and 600C. In a similar study of laser-deposited and sintered (1000C, 6h) orthorhombic $Sc_2W_3O_{12}$, the average thermal expansion coefficient was -5.28 x 10^{-6}/K between room temperature and 600C. The as-deposited thin films were again amorphous. The surface of the as-deposited film became uneven as the substrate temperature was increased[183]. Following annealing (1000C, 420s), the thermal expansion coefficient was -7.17 x 10^{-6}/K between room temperature and 600C.

A full analysis of thermal motion has been carried out for $Sc_2(WO_4)_3$ between 50 and 823K. By using neutron diffraction data from isotopically pure samples of $Sc_2(^{184}WO_4)_3$, $Sc_2(^{186}WO_4)_3$ and natural $Sc_2(WO_4)_3$, it was possible to extract anisotropic thermal parameters for individual atoms over the above temperature range. This revealed that the thermal motion of two Sc-O-W bridging oxygen atoms with the largest Sc-O-W angles is better represented by thermal toroids, consistent with strong local motion of the units. The thermal behavior of the other oxygen atoms in the structures, as a function of temperature, is normal[184]. The Sc-O and W-O bond lengths, corrected for the effect of correlated thermal motion, exhibit the expected increase with temperature.

Pressure-induced amorphization in scandium compounds which exhibit negative thermal expansion is such that a change in electrical resistance and structure occurs with increasing pressure[185]. Experimental analysis of pressure-induced disordering has indicated that heating and structural collapse of the material is related to the densification mechanism. The relationship between polyatomic anion conduction and negative thermal expansion in the $Sc_2(WO_4)_3$-type structure has been investigated. The motion of the effective charge carriers can be visualised using molecular dynamics simulation if the initial structure and force-field are known. By reproducing the negative thermal expansion over a wide temperature range, a force-field can be deduced which predicts the mobile species. By using the same force-field, a series of correlated WO_4^{2-} migrations is observed in extended isothermal-isobaric simulations. Tubandt-type electrolysis experiments confirm that the mobile species is anionic[186]. Scandium tungstate is therefore the prototype of a class of WO_4^{2-} anion conductors.

In order to study the relationship between structural characteristics and phonon properties in negative thermal expansion compounds, the heat capacities of $Sc_2W_3O_{12}$ and $Sc_2Mo_3O_{12}$ were measured. Spectral analysis of the heat capacity furnished the effective phonon density of states. That of $Sc_2W_3O_{12}$ had three features: a low-energy phonon mode with negative mode-Grüneisen parameter of about 5meV, high-energy phonon modes and a separation of the phonon density-of-states into two regions with a wide separation. The low-energy phonon modes with negative Grüneisen parameter caused the negative thermal expansion, and the other two features were necessary in order to maintain the negative expansion over a wide temperature range[187]. A comparison of the phonon density-of-states with that of other oxides indicated that such phonon features are common in negative thermal expansion oxides and are related to their common chemical and structural characteristics of strong bonding and a framework structure. A composite in which a core of $Sc_2W_3O_{12}$ was coated with copper using simple electroless plating methods has been studied. Copper nanocrystals first formed, and then grew together to produce a continuous coating[188]. The $Sc_2W_3O_{12}$/Cu composite had a linear coefficient of thermal expansion of -4.47 x 10^{-6}/K between room temperature and 200C.

Solid solutions of the form, $Sc_{2-x}Ga_xW_3O_{12}$ (x = 0, 0.05, 0.1, 0.2, 0.3, 0.5, 0.8), were prepared by solid-state reaction at 1100C. All of the samples exhibited negative thermal expansion between 25 and 1000C. The unit-cell parameters, and cell volume, all decreased with increasing gallium content apart from the b-axis, which exhibited expansion. The average volume expansion coefficient also decreased with increasing gallium content[189]. As the temperature increased, the value of the volume expansion coefficient decreased markedly between room temperature and 300C, but remained almost unchanged between 300 and 800C. It decreased further at temperatures greater than 800C, tended to zero and then became positive.

$Y_2(WO_4)_3$

Early neutron diffraction studies of polycrystalline $Y_2W_3O_{12}$ at 15 to 1373K showed that all three cell-edges of the orthorhombic phase decreased with increasing temperature, giving an average linear thermal expansion coefficient of -7.0 x 10^{-6}/K. Rietveld refinement of the data revealed an apparent 0.05Å decrease in the average W-O distance between 15 and 1373K. This apparent decrease caused the decrease in the cell edges, but it was not a real decrease in the average W-O distances[190]. The apparent shrinkage was instead attributed to the transverse thermal motion of oxygen in the Y-O-W linkages.

Thin films of $Y_2W_3O_{12}$ have been grown onto quartz substrates by means of pulsed laser deposition, and the negative thermal expansion investigated using high-temperature X-ray diffraction. All of the as-deposited films were amorphous, regardless of the oxygen pressures of 5 to 20Pa and substrate temperatures of 298 to 500C. The films which were grown at 500C under an oxygen pressure of 10Pa had the smoothest and most uniform surface morphology. Crystalline orthorhombic thin films were prepared by heating (1000C, 420s, air). The crystallized film was polycrystalline and had a compact surface morphology. The mean grain size was between 0.2 and 1μm. The orthorhombic thin film exhibited anisotropic negative thermal expansion. The thermal expansion coefficients for the a-axis, b-axis c-axis and volume were -12.33 x 10^{-6}, -4.99 x 10^{-6}, -11.01 x 10^{-6} and -28.22 x 10^{-6}/K, respectively, between 100 and 600C[191]. The average linear thermal expansion coefficient was -9.41 x 10^{-6}/K. When powder samples of Y_2O_3 and WO_3 were planetary ball-milled, differential thermal analysis suggested that the reaction temperature of the as-mixed powder was 1000C. The average particle size was about 65nm and $Y_2W_3O_{12}$ formed at 800C. Low-temperature synthesis retained a finer grain size and improved densification and sinterability[192]. The resultant material had a thermal expansion coefficient of the order of -7.1 x 10^{-6}/K at 150 to 650C. Yttrium tungstate was synthesized via rapid solidification using a CO_2 laser. The orthorhombic densely-packed layer structure could change from $Y_2W_3O_{12}$ to $Y_2W_3O_{12}•xH_2O$ by adsorbing water from the air when stored at room atmosphere. Raman spectra revealed that the presence of water molecules affected the rocking, and every other type of motion, of the corner-shared polyhedra. Water molecules that could be released below 388K had little effect upon the vibrations of the [WO_4] tetrahedra. Those which could be released at above that temperature not only hindered rocking, but also stretching, bending, librational and translational motions of the polyhedra[193]. The temperature dependence of the Raman spectra indicated that not only low-frequency modes but also higher optical phonon modes contributed to the negative thermal expansion. Negative thermal expansion in the orthorhombic tungstate was measured using high-temperature X-ray diffraction[194]. The coefficients of thermal expansion were -9.78 x 10^{-6}, -5.13 x 10^{-6}, -6.68 x 10^{-6} and -22.02 x 10^{-6}/K for the a-axis, b-axis, c-axis and volume, respectively, between 473 and 1073K. Neutron powder diffraction experiments, performed between 20 and 800C, indicated coefficients of thermal expansion of -10.35 x 10^{-6}, -3.06 x 10^{-6} and -7.62 x 10^{-6}/K for the a-axis, b-axis and c-axis, respectively[195].

A simulation of the negative thermal expansion was performed by making calculations of phonon dispersion curves using density functional perturbation theory. The mode eigenvectors were mapped onto flexibility models, and the results were compared with calculations of the mode Grüneisen parameters[196]. Many lower-frequency phonons were found to contribute to negative thermal expansion in $Y_2W_3O_{12}$, all of which could be described in terms of rotations of rigid WO_4 tetrahedra and Y-O rods. A combination of classical and first-principles density functional theory methods has been used to study the lattice dynamics of $Y_2W_3O_{12}$. The Born dynamic charges of various atoms were found to deviate anomalously from the nominal values. Phonons with energies ranging from 4 to 10meV are the main contributors to the negative thermal expansion[197]. Those phonons involve rotations of the YO_6 octahedra and WO_4 tetrahedra in mutually opposite senses, plus collective translational atomic displacements.

Solid solutions of the form, $Y_{2-x}La_xW_3O_{12}$, were synthesized by solid-state reaction. The structure changed from orthorhombic to monoclinic with increasing lanthanum substitution and a high lanthanum content led to a low relative density. The material was orthorhombic for $0 \leq x \leq 0.4$ and monoclinic for $1.5 \leq x \leq 2$. The thermal expansion coefficients for $0 \leq x \leq 2$ varied from -9.59 x 10^{-6} to 2.06 x 10^{-6}/K with increasing lanthanum content[198]. The composition, $Y_{0.25}La_{1.75}W_3O_{12}$, exhibited almost zero thermal expansion and its average linear thermal expansion coefficient was -0.66 x 10^{-6}/K between 103 and 700C. Solid solutions of the form, $Y_xNd_{2-x}W_3O_{12}$ (x = 0.0 to 1.0 and 1.6 to 2.0), were synthesized using solid-state methods. The material was monoclinic for x = 0.0 to 1.0 and orthorhombic for x = 1.6 to 2.0. The thermal properties of $Y_xNd_{2-x}W_3O_{12}$ (x = 0.25 and 1.90) depended strongly upon both their structures and compositions. The positive thermal expansion of $Y_{0.25}Nd_{1.75}W_3O_{12}$ was anisotropic along the three crystallographic axes, with the a-axis expanding between 25 and 200C but contracting between 200 and 800C. The b- and c-axes expanded over the entire temperature range[199]. The thermal expansion coefficient of $Y_{1.9}Nd_{0.1}W_3O_{12}$ was -17.9 x 10^{-6}/K between 200 and 1000C; compared to -20.9 x 10^{-6}/K for $Y_2W_3O_{12}$.

When solid solutions of the form, $Y_2W_{3-x}Mo_xO_{12}$ ($0.5 \leq x \leq 2.5$), were synthesized using solid-state methods, they all had an orthorhombic structure and exhibited a negative thermal expansion which was related to the transverse vibration of bridging oxygen atoms. The thermal expansion coefficient for $Y_2W_{3-x}Mo_xO_{12}$ was 16.2 x 10^{-6}/K for x = 0.5, and -16.5 x 10^{-6}/K for x = 2.5, between 200 and 800C[200]. High-temperature X-ray diffraction data suggested that the difference between W-O and Mo-O bonds was responsible for the change in thermal expansion coefficients caused by element substitution.

$Yb_2(WO_4)_3$

Negative thermal expansion in the orthorhombic tungstate was measured by using high-temperature X-ray diffraction[201]. The coefficients of thermal expansion were -10.20 x 10^{-6}, -2.65 x 10^{-6}, -6.41 x 10^{-6} and -19.14 x 10^{-6}/K for the a-axis, b-axis, c-axis and volume, respectively, between 473 and 1073K. Solid solutions of the form, $Yb_2W_{3-x}Mo_xO_{12}$ ($0.5 \leq x \leq 2.5$), were prepared by using high-temperature solid-state methods. All of the samples had the same orthorhombic structure and exhibited a negative thermal expansion which was related to transverse vibration of the bridging oxygen atoms in the structure. The thermal expansion coefficients were -18.36 x 10^{-6}/K for x = 0:5 and -19.6 x 10^{-6}/K for x = 2.5; as compared to -19.1 x 10^{-6}/K and -18.3 x 10^{-6}/K for unsubstituted $Yb_2W_3O_{12}$ and $Yb_2Mo_3O_{12}$, respectively, between 200 and 800C[202]. High-temperature X-ray diffraction data suggested that the difference between the W-O and Mo-O bonds was responsible for the change in thermal expansion coefficients following element substitution.

$(Yb,La)_2(WO_4)_3$

In solid solutions of the form, $Yb_{2-x}La_xW_3O_{12}$, the structure changes from orthorhombic to monoclinic with increasing lanthanum substitution. Pure phases can form with an orthorhombic structure when $0 \leq x \leq 0.5$ and with a monoclinic one when $1.5 \leq x \leq 2$. A high lanthanum content also leads to low hygroscopic tendencies[203]. The thermal expansion coefficients of samples with $0 \leq x \leq 2$ varies from -7.78 x 10^{-6} to 2.06 x 10^{-6}/K across that range of lanthanum contents.

$Zr(WO_4)_2$

General

Negative thermal expansion was found for ZrW_2O_8, from 0.3K to its decomposition temperature of about 1050K. Cubic symmetry persists over its entire stability range and the negative thermal expansion behavior is isotropic[204]. No other material was then known to exhibit such a behavior over such a wide temperature range. The negative thermal expansion is not disrupted by the structural phase transition at 430K.

X-ray absorption fine structure results reveal a very small temperature dependence of the broadening parameter for the W-Zr atom pair and the W-O-Zr linkage; suggesting that the displacements of the tungsten, oxygen and zirconium atoms must be correlated. A much larger temperature dependence of the broadening parameter is found for the nearest W1-W2 pair as well as for the nearest Zr-Zr pair. The combined results suggest that the correlated motion of a WO_4 tetrahedron and its three nearest ZrO_6 octahedra leads to the negative expansion effect rather than the transverse vibrations of the oxygen atom in the

W-O-Zr linkage. The data for W-W and Zr-Zr pairs also suggest hardening of the spring constant near to 100K[205]. This is consistent with the shift in the lowest mode, as a function of temperature, in the phonon density of states.

Below 428K, it has a $P2_13$ well-ordered structure containing corner-sharing ZrO_6 octahedra and two crystallographically distinct WO_4 tetrahedra (figure 3). Above the apparently second-order phase transition at 428K, the space group is Pa3. The structure is then disordered, with one oxygen site being 50% occupied and thus suggesting high oxygen mobility[206]. Oxygen motion above 428K is also suggested by dielectric and alternating-current impedance measurements. A complicated distribution in reciprocal space of rigid unit mode phonons has been identified. These are very low energy phonons which cause collective rotations of the ZrO_6 octahedra and WO_4 tetrahedra within the crystal structure[207]. These distortions then permit ZrW_2O_8 to contract upon heating. Negative thermal expansion is now recognized in a large new family of materials having the general formula, $A_2(MO_4)_3$[208]. Substitution markedly affects the thermal expansion properties, thus permitting the tailoring of ceramics with negative, positive or zero coefficients of thermal expansion. The surfaces of these modes in wave-vector space have been located.

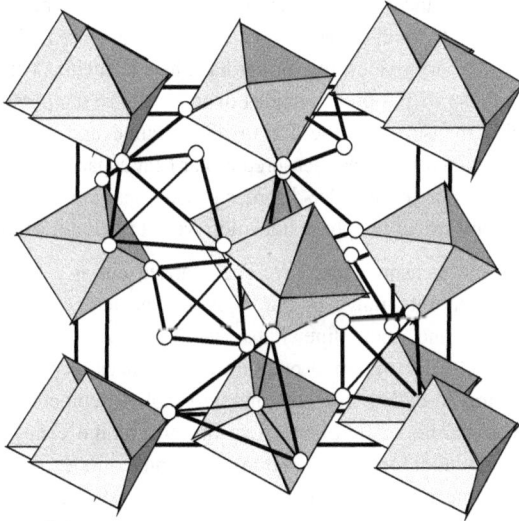

Figure 3 Structure of Zirconium Tungstate. The octahedra are depicted as solids and the tetrahedra as wire frames. Only the tetrahedron atoms are shown.

The rigid-unit mode interpretation accounts for the weak effect of the 430K structural phase transition upon the negative thermal expansion; provided that the disordered phase does not involve the formation of W_2O_7 and W_2O_9 complexes. The crystal structure of the related material, ZrV_2O_7, is cross-braced by pairs of linked tetrahedra and is consequently much less flexible[209]. A qualitatively different mechanism could be responsible for the negative thermal expansion of ZrV_2O_7. Detailed lattice dynamics calculations of the large isotropic negative thermal expansion at 0 to 1050K reflected the unusually dominant quantitative contributions made to the negative expansion by the transverse acoustic, and librational and translational optical, phonons below 8meV[210]. These contributions are reduced in the high-pressure orthorhombic phase, which thus exhibits negative expansion only below 270K. The low-energy modes causing negative thermal expansion, and corresponding to the correlated vibrations of a WO_4 tetrahedron and its three nearest ZrO_6 octahedra, are revealed by X-ray absorption fine structure data[211]. An anisotropic soft mode arising from the so-called frustrated soft mode geometrical phenomenon could be prevented by the interconnectivity of the modes. Experimental heat capacity measurements of α-ZrW_2O_8 were made from 0.6 to 400K, and measurements of β-$ZrMo_2O_8$ from 8 to 400K[212]. These confirmed the existence of very low frequency modes in both materials. At 298.15K the standard molar heat capacity of the α-ZrW_2O_8 was 207.01J/Kmol. The standard molar entropy was 257.96J/Kmol[213]. When the heat capacities of negative thermal expansion materials are compared with the weighted sums of their constituent binary oxides, the former have heat capacities which are much greater than the sum of those of the binary oxides over the entire temperature range. The heat capacity was also measured in the 1.8 to 483K range and standard thermodynamic functions were estimated from the measured molar heat capacity. A large λ-type anomaly of the heat capacity, due to an order-disorder phase transition, was observed at 440K[214]. The enthalpy and entropy of the phase transition were 1560J/mol and 4.09J/Kmol.

Raman spectroscopic measurements have been carried out as a function of temperature over the complete stability range of the α-phase. The temperature dependence of the phonon frequencies and their linewidths were analyzed on the basis of known anharmonicity models. The reported pressure dependences of phonon frequencies were used to separate the so-called true anharmonic (explicit temperature-effect at constant volume) and quasi-harmonic (implicit effect due to volume change) parts of the total anharmonicity[215]. In addition to the 41/cm librational mode, some other lattice and bending modes of the tungstate ion were found to be strongly anharmonic. *Ab initio* density functional theory calculations have been made of phonons in the cubic phase of ZrW_2O_8 over the entire Brillouin zone. Minimal models have been used to determine the origin of large negative thermal expansions in under-constrained systems. Under-constraint can organize a thermodynamically extensive manifold of low-energy modes which drives the negative thermal expansion but extends across the Brillouin zone[216]. Mixing of twist and translation in the eigenvectors of these modes for which, in the case of ZrW_2O_8, there is infra-red and neutron-scattering evidence, becomes a signature of the dynamics of under-constraint.

It was found that those phonons exhibiting anharmonicity with temperature were not necessarily the same ones as those exhibiting anharmonicity with pressure; although both had similar frequencies. Only the latter phonons were associated with negative thermal expansion, and the cubic and/or quadratic anharmonicity of phonons is not relevant to negative expansion; only the volume dependence of the frequencies[217]. These calculations could reproduce observed anomalous trends, such as softening with pressure of the low-frequency peak at about 4meV in the phonon spectra and its hardening with temperature. Phonon dispersion measurements of a single crystal of ZrW_2O_8 verified the prediction of highly anharmonic specific phonon branches such as the transverse acoustic branch, and other branches up to 10meV[218]. The modes below 10meV mainly contributed to the negative thermal expansion.

Perturbed angular correlation studies reveal the occurrence of four distinct zirconium sites and, based upon the values of the quadrupole parameters, it is deduced that one site is associated with normal ZrO_6 while the other three sites are associated with contracted and distorted ZrO_6. An effective decrease in the Zr-O bond length is attributed to an increase in the fraction of those ZrO_6 octahedra which undergo contraction with temperature; thus explaining the observed overall negative thermal expansion[219].

The local structure of the low-temperature ordered phase of ZrW_2O_8 was investigated via reverse Monte Carlo modelling of neutron total scattering data. Quantitative measurements were obtained, for the first time, of the extent to which the WO_4 and ZrO_6 polyhedra move as rigid units. These values were moreover consistent with the predictions of rigid unit mode theory[220]. The results did not support a previous interpretation of X-ray absorption fine structure spectroscopy data in terms of a larger rigid structural component involving the Zr-O-W linkage. In fact, in spite of ZrW_2O_8 being the most studied negative thermal expansion material, the nature of the exact mechanism involved is still not clear. The so-called tent-model is a rival to the rigid-unit-mode model. Molecular dynamics simulations have been used to distinguish, for each bond distance, the so-called true thermal expansion from the apparent thermal expansion. A decrease in the true W-Zr distances was observed, accompanied by large transverse vibrations of the O atoms at the middle of W-O-Zr linkages. This differed greatly from the tent-model view. Contrary to the rigid-unit-mode model, the WO_4 and ZrO_6 polyhedra are strongly distorted by thermal motion, and intra-polyhedra contributions to negative expansion occur[221]. On this basis, both models are unable to explain properly the negative thermal expansion of ZrW_2O_8. A more flexible model, based upon just rigid nearest W-O and Zr-O bonds and tension effects, has been suggested. An X-ray pair distribution function and extended X-ray absorption fine-structure study of ZrW_2O_8 at 10 to 500K had yielded information on the stiffness of the Zr-O-W linkage. The former technique was highly sensitive to W-Zr and W-W correlations, but much less so to O-O or W-O correlations. The Zr-W peak in the pair-distribution data exhibited a weak temperature dependence. This linkage is relatively stiff and does not permit bending of the Zr-O-W. It was suggested that the low-energy vibrational modes that lead to negative thermal expansion involve correlated rotations of ZrO_6 octahedra that produce large [111] translations of the WO_4 tetrahedra, rather than transverse motions of oxygen atoms which would imply a flexible Zr-O-W linkage[222].

The low-frequency phonon modes of cubic ZrW_2O_8 and HfW_2O_8, and of trigonal $ZrMo_2O_8$ and $HfMo_2O_8$ were compared using Raman and terahertz time-domain spectroscopy. A number of distinctive low-frequency modes appeared below 150/cm in the ZrW_2O_8 and HfW_2O_8 and were attributed to the librational and translational motions of polyhedra. Only one weak mode was present in the Raman spectra, but absent from the THz spectra, of trigonal $ZrMo_2O_8$ and $HfMo_2O_8$ and was attributed to interlayer breathing. The lowest optical phonon mode, at about 40/cm, disappeared and this was associated with obvious weakening of the lowest asymmetrical stretching-mode for ZrW_2O_8 and HfW_2O_8 across the order-disorder phase transition. This was direct evidence of a reduction in the number of rigid unit modes in the high-temperature phase. The correlated librational and translational motions of the WO_4 tetrahedra and ZrO_6/HfO_6 octahedra with out-of-phase asymmetrical stretching of the two neighboring WO_4 tetrahedra, explained a large fraction of the negative thermal expansion in the low-temperature phase. Some of the correlated motions could be destroyed across the order-disorder phase transition, thus causing a smaller negative thermal expansion in the high-temperature phase[223]. The lack of rigid unit modes in trigonal $ZrMo_2O_8$ and $HfMo_2O_8$ is the cause of their positive thermal expansion. Low-energy phonon dynamics in ZrW_2O_8 were investigated using infrared spectroscopy between 5000 and 16/cm. It was suggested that the prevalence of infrared active phonons at low energy, and their observed temperature dependence, were highly unusual and indicative of exotic low-energy lattice dynamics[224]. It was also suggested that the lowest optical modes in ZrW_2O_8 tended to have a mixed librational and translational nature, in which the unconstrained oxygen played an essential role.

Negative thermal expansion materials, such as ZrW_2O_8 and $HfMo_2O_8$, also have very low thermal conductivities[225]. This has been attributed to efficient coupling of the low-frequency optical phonons, which give rise to the negative thermal expansion, to the heat-conveying acoustic phonons. In order to investigate the relationship between negative thermal expansion and other thermal properties, the thermal conductivity of the α-phase of ZrW_2O_8 has been determined from 1.9 to 390K[226]. The conductivity is glass-like, and close to the theoretical minimum value.

The covalent bonding nature of ZrW_2O_8 was revealed by accurate charge-density distribution determinations at 300K. Two WO_4 tetrahedra are bonded covalently by a bridging O atom along the body diagonal directions of the cubic structure[227]. The bonding electron density between the tungsten atom and the terminal oxygen atom is highest among the W-O covalent bonds.

The specific heat of ZrW_2O_8, measured from 1 to 300K, indicates that - below 50K – it is dominated by an Einstein energy term of about 5meV; representing 6.5% of the total lattice degrees of freedom[228]. This concentration of phonon spectral weight at low energies is unusual in cubic materials and is a possible indication of the existence of the above rigid unit modes. Accurate lattice parameters were determined between 2 and 520K. Two different analyses of the temperature dependence of the lattice constant provided direct evidence that the negative thermal expansion behaviour is due to low-frequency modes on an energy scale of 3 to 8meV[229]. These rigid-unit modes are associated with very large Grüneisen parameters. The Grüneisen parameters of phonon modes have been determined as a function of their energy by making high-pressure inelastic neutron scattering measurements[230]. The large negative thermal expansion was deemed to be due to a marked softening of the phonon spectrum at 1.7kbar. A negative Grüneisen parameter mode is consistent with the observed Raman spectral peak of 40/cm[231]. From equation-of-state considerations, negative thermal expansion can be related to a negative (thermal or electronic) Grüneisen parameter. Under pressure, this leads to equation-of-state anomalies, with the pressure-derivative of the bulk modulus becoming small or negative. Many such materials exhibit a pressure-induced amorphization which can be understood in terms of steric constraint. Negative thermal expansion can usually be traced to the existence of two degenerate, or almost degenerate, energy states[232]. Upon increasing the temperature, the material then favours the lower volume state.

Mode Grüneisen parameters have been estimated for α-ZrW_2O_8 zone-center modes by means of density functional theory calculations. The temperature-dependence of the coefficient of thermal expansion was deduced from the Debye-Einstein model in the quasi-harmonic approximation. The lowest-energy optical modes were found to be at 45 and 46/cm, and were shown to be mainly responsible for the negative thermal expansion at low temperatures. Experimental evidence for the lowest-energy triply-degenerate infrared active optical mode was found in the far infrared spectrum[233]. With increasing temperature, other optical modes, at 96, 100, 133, 161 and 164/cm also contributed to the thermal expansion near to room temperature.

High-pressure X-ray diffraction and Raman scattering experiments showed that ZrW_2O_8 becomes progressively amorphous from 1.5 to 3.5GPa. The amorphous phase was retained after releasing the pressure, but the original crystalline phase returned upon annealing (923K)[234]. These results indicated a general trend between negative thermal expansion and pressure-induced amorphization in highly flexible framework structures. Measurements of the elastic constants of ZrW_2O_8 reveal relatively soft and nearly isotropic elastic constants, with normal Poisson ratios, plus normal ambient-temperature elastic and dominant low-frequency acoustic vibration modes[235].

The negative thermal expansion has been investigated at the microscopic scale by measuring electric-field gradients at ^{187}W nuclear probes. Two distinct nuclear quadrupole interactions were observed at 295K and attributed to the two crystallographically distinct WO_4 tetrahedra of the room-temperature structure. The observed temperature dependence of the nuclear quadrupole interactions agreed with the structural phase transition at 428K[236]. These experiments supported the mechanism of coupled librations of rigid ZrO_6 octahedra and WO_4 tetrahedra.

High-resolution powder diffraction data on cubic ZrW_2O_8 [a = 9.18000Å at 2K], for 260 temperatures between 2 and 520K, confirmed a negative coefficient of thermal expansion of -9.07 x 10^{-6}/K. The phase transition could be described in terms of a simple cubic three-dimensional Ising model, but with unusual associated kinetics[237]. A hysteresis of the cell parameter through the phase transition was the opposite of that normally observed.

When differential scanning calorimetry measurements were performed in dry N_2, an endothermic peak and base-line shift were observed, at 124 and 169C respectively, upon heating as-prepared ZrW_2O_8. During successive cooling and heating measurements, the base-line shift at 169C persisted but the peak at 124C disappeared. The base-line shift at 169C corresponded to a variation in the thermal expansion coefficient due to a structural phase transition from an acentric cubic phase to a centric phase[238]. Simultaneous thermogravimetry and thermogravimetry differential thermal analysis mass spectroscopy measurements revealed that the endothermic peak at 124C was due to secession of the H_2O (less than 1.6mol%) involved with the ZrW_2O_8. This trace amount negligibly affected the thermal expansion behavior, and the variation in the thermal expansion coefficient at around 164C was attributed to the λ-type transition.

Synthesis

An early method used for the synthesis of zirconium tungstate was based upon spray-drying. The starting materials were ammonium metatungstate, zirconium oxychloride and zirconium oxynitrate; chosen for their high solubilities in water. Spray drying of the slurry gave a powder of increased reactivity due to good intermixing of the Zr- and W-phases. Following pre-heating (700C, 10h), calcination (1180C, 1h) in sealed quartz tubes produced high-quality ZrW_2O_8. As compared with samples prepared using solid-state reaction between ZrO_2 and WO_3, the spray-dried product had a very high density and higher purity.[239] Substrates of ZrW_2O_8 could be formed by sintering (1155 to 1200C, 6h). The thermal expansion coefficient, Young's modulus and apparent density increased with increasing sintering temperature[240]. The enthalpy of formation of ZrW_2O_8 has been estimated from the enthalpies of changes in the oxygen coordination of cations upon the formation of mixed oxides from binary oxides[241]. The formation of the tungstate from the constituent oxides is accompanied by heat absorption. When ZrW_2O_8 was synthesized via solid-state reaction, using analytically pure ZrO_2 and WO_3, the average thermal expansion coefficient of the cubic ZrW_2O_8 was -5.605 x 10^{-6}/K between room temperature and 600C[242]. That of sintered ceramic rod was -10.709 x 10^{-6}/K and that of pressed powder rod was -4.490 x 10^{-6}/K.

When ZrW_2O_8 was synthesized by solid-state reaction, samples that were sintered (1220C, 3h) comprised a single cubic phase of ZrW_2O_8 with an average dimension of 0.5µm. The thermal expansion coefficient was -11.58 x 10^{-6}/K between ambient and 150C and -3.77 x 10^{-6}/K from 200C to 500C; with an average value of -6.31 x 10^{-6}/K.[243] When ZrW_2O_8 was synthesized from ZrO_2 and WO_3 by reactive sintering (1200C, 24h), the thermal expansion coefficient between 18 and 718C was -5.57 x 10^{-6}/K[244]. The density was 4.81g/cm^3.

Powders of ZrW_2O_8 have also been synthesized by co-precipitation, using polyethylene glycol as a dispersant to control the particle size. The samples were single-phase α-ZrW_2O_8 of regular shape. The thermal expansion coefficient was -10.35 x 10^{-6}/K from room temperature to 150C and -3.08 x 10^{-6}/K from 200 to 600C; with an average value of -5.38 x 10^{-6}/K[245]. Ultrafine cubic material, obtained by solid-state reaction of nanometer ZrO_2 and WO_3 powders had an average thermal coefficient of -6.82 x 10^{-6}/K between 20 and 600C.[246]

Upon comparing the effect of these various preparation methods, it was found that the powders produced by co-precipitation were the largest, with an average dimension of 2.5µm x 3.0µm, while powders prepared by a sol-gel method were the smallest, with an average dimension of 100nm. The thermal expansion coefficient was about the same when the particle size was in the micron range[247]. When the particle size was in the nanometer range, the thermal expansion coefficient decreased with particle size. The ZrW_2O_8 has also been synthesized from $ZrOCl_2 \cdot 8H_2O$ and H_2WO_4 using co-precipitation. A suitable pH value for the co-precipitation of Zr^{4+} and WO_4^{2-} was 2 to 3. After heating (1200C, 1h), the main phase of the synthesized powders was ZrW_2O_8[248]. The particle size was about 100nm and was homogeneously distributed.

A hydrothermal method, using sodium dodecyl benzene sulfonate as a surfactant, produced single cubic phase powders of high crystallinity. With increasing amounts of surfactant, the morphology and particle size changed from rod-like, with average dimensions of $1.2\mu m$ x $1.2\mu m$ x $10\mu m$, to layered columns with an average diameter of $0.4\mu m$[249]. The amount of surfactant did not affect the average thermal expansion coefficient. The latter were all similar, with an average value of -5.8 x 10^{-6}/K between room temperature and 500C[250]. On the other hand, the morphology had some effect upon the thermal expansion coefficient.

The material can be easily obtained from the hydrated precursor, $ZrW_2O_7(OH)_2 \cdot 2H_2O$. Optimum processing results in rod-shaped particles with widths of 10 to 30nm and lengths of 200 to 500nm; with little agglomeration[251]. The particle size and morphology of the precursor are preserved during conversion to the desired phase at 600 to 650C.

Rapid quenching of ZrW_2O_8 melt from 1300C gave dense specimens with few inner pores. Dilatometry revealed a sudden thermal expansion at 120 to 160C, which was attributed to the removal of internal stresses that had built up during heating from room temperature to 300C. Annealing (200C) to remove the internal stress led to a negative average thermal expansion coefficient of some -4.5 x 10^{-6}/K between room temperature and 250C. This was smaller than that (-8 x 10^{-6}/K) of sintered material[252]. The present specimens were a mixture of ZrO_2 and/or WO_3 grains of about $1\mu m$ diameter, and ZrW_2O_8 grains with a diameter of 10 to $25\mu m$. The lower negative thermal expansion was attributed to decomposition of the ZrW_2O_8 during quenching.

When prepared from ZrO_2 and WO_3 powders by firing (1200C, air) followed by cooling in a furnace, in air, in water or in liquid nitrogen, samples which were furnace-cooled exhibited a positive thermal expansion while quenched samples exhibited an expansion of about -0.004/K from room temperature to 600C[253].

Nanoparticles of ZrW_2O_8 having various morphologies were synthesized by using sol-gel and hydrothermal methods and the thermal expansion was studied between 25 and 600C. In the case of the α-phase, the coefficients were -12.0 x 10^{-6} and -13.2 x 10^{-6}/K for sol-gel and hydrothermal samples, respectively. In the case of the β-phase, the coefficients were -4.3 x 10^{-6} and -5.1 x 10^{-6}/K for sol-gel and hydrothermal samples, respectively[254]. The α-phase coefficients for nanoparticles made using either method were higher than that (-8.9

x 10^{-6}/K) for bulk ZrW_2O_8. When synthesized using hydrothermal methods, and heat-treated (570C, 6h), the powder was single-phase α-ZrW_2O_8 with a regular rectangular particle shape and an average size of 1.2μm x 1.2μm x 10μm[255]. The thermal expansion coefficient was -6.30 x 10^{-6}/K between room temperature and 500C. The α-β transition temperature was between 150 and 175C. Samples were synthesized by hydrothermal methods, with low-temperature (500C) heat treatment[256]. The crystalline precursor, $ZrW_2O_7(OH)_2 \cdot 2H_2O$ was formed when the HCl concentration was at least 6mol/L. When samples were synthesized using the sol-gel method, with citric acid as a chelating agent, the results were single-phase α-ZrW_2O_8 of irregular shape. With increasing citric acid content, the particles grew larger and layered[257]. From room temperature to 500C, the average thermal expansion coefficient was -6.14 x 10^{-6}/K. Thin films synthesized using the sol-gel method were single-phase α-ZrW_2O_8 and the surfaces had small holes with no obvious cracks. The film thickness was about 500nm[258]. The film was transparent in the visible region, with two absorption peaks in the ultraviolet region at 348.3 and 277.2nm. Powders were synthesized using sol-gel methods and compared with powders synthesized by solid-state reaction. The average dimensions were smaller in the former samples; the average size being 100nm[259]. The thermal expansion coefficient of material synthesized by sol-gel chemistry was -5.93 x 10^{-6}/K between room temperature and 500C; lower than that (-6.31 x 10^{-6}/K) for solid-state reaction. Various acids were used to prepare ZrW_2O_8 at 570C by using the hydrothermal method. Among the chosen acids, only HCl and HNO_3 could be used to prepare pure powders of high crystallinity and rod-like shape. When the hydrothermal temperature was reduced to 160C, nano-sphere particles with an average diameter of 30nm were obtained, if prepared by HNO_3 addition[260]. The powders prepared using pure HCl or HNO_3 exhibited a marked negative thermal expansion. Nanorods of ZrW_2O_8 were produced by using the hydrothermal method, followed by post-annealing (570C, 2h). The precursor, $ZrW_2O_7(OH) \cdot 2(H_2O)_2$, depended significantly upon the HCl concentration and could form in 2 to 8mol/L HCl solution[261]. Upon increasing the concentration from 2 to 8mol/L, the rod-like particles became more homogeneous and the average dimensions changed from 10μm x 0.5μm to 700nm x 50nm. All of the ZrW_2O_8, obtained under various conditions, exhibited negative thermal expansion and the average coefficients, between 15 and 600C, gradually decreased with increasing HCl concentration.

Isotropic ZrW_2O_8 powders were synthesized by combustion at low temperatures. High-purity powders with a mean particle size of 0.5μm could be obtained by using a furnace temperature of 500C, a H_3BO_3 molar fraction of 10%, a 2:1 mass ratio of $(NH_2)_2CO$ and $(NH_4)_5H_5$ $[H_2(WO_4)_6] \cdot H_2O$ added to $ZrOCl_2 \cdot 8H_2O$ and a 1:3.2 molar ratio of $(NH_4)_5H_5[H_2(WO_4)_6] \cdot H_2O$ and $ZrOCl_2 \cdot 8H_2O$[262]. The linear thermal expansion coefficient was -5.08 x 10^{-6}/K at 50 to 700C, and the temperature-dependence was described by - 0.014 - 4.5 x 10^{-4}T.

Thin films have been grown onto quartz substrates by using pulsed laser deposition. The as-deposited film contained an amorphous phase and the stoichiometry of the as-deposited film was close to that of the ZrW_2O_8 ceramic target. Crystallized cubic ZrW_2O_8 thin films were prepared by annealing (1200C). The thin film deposited onto a substrate at 650C was smooth and compact. The crystallized cubic ZrW_2O_8 thin film was polycrystalline[263]. The thermal expansion coefficient was -11.378 x 10^{-6}/K between 20 and 600C. Thin films have also been deposited onto quartz substrates by means of alternating radio-frequency magnetron sputtering with WO_3 and ZrO_2 ceramic targets. The as-deposited film was again amorphous and its surface was smooth and granular[264]. Cubic ZrW_2O_8 thin film was prepared by annealing (1200C, 180s). The cohesion between the quartz substrate and the ZrW_2O_8 film was good. The thermal expansion coefficient of the film was -24.81 x 10^{-6}/K at 15C to 200C, and -4.78 x 10^{-6}/K at 200C to 700C[265]. The average coefficient was -10.08 x 10^{-6}/K between 15 and 700C. Similar studies have shown[266], on the basis of variations in the interplanar distances of the (211) crystal plane at various temperatures, that the coefficient of thermal expansion of ZrW_2O_8 thin films is -25.4 x 10^{-6}/K.

Films have been deposited using electron beam evaporation and reactive co-sputtering. The evaporated films, but not the denser sputtered films, exhibited negative thermal expansion. The negative expansion behavior occurred for a range of stoichiometries. Because crystalline ZrW_2O_8 and thin-film ZrW_xO_y both have low densities and both exhibit negative expansion, it was suggested that similar mechanisms are operating[267]. In addition, because the thin-film deposition conditions can often be used to control the density, it was speculated that more thin-film, than bulk, materials could be caused to exhibit negative thermal expansion.

Effect of Additions

Doubly-substituted Al^{3+}, V^{5+} solid solutions, $Zr_{1-x}Al_xMoW_{1-y}V_yO_8$, were prepared by co-precipitation. The density increased with increasing V^{5+} content while its porosity declined[268]. The thermal expansion coefficients were between -3.0 x 10^{-6} and -4.0 x 10^{-6}/K; indicating that the negative thermal expansion behavior was slightly inferior to that of $ZrMoWO_8$. The samples also exhibited considerable heat-shock resistance.

The thermodynamic properties of ZrW_2O_8, when substituted at the zirconium site, were studied from the viewpoint of the ionic radii of the substituted and host ions and of oxygen defects. The entropies of the α-to-β phase transitions in the $Zr_{1-x}Hf_xW_2O_8$ series (x = 0, 0.5, 1) were all the same (2.1J/Kmol). The normalized (310) diffraction peaks of

$Zr_{1-x}Hf_xW_2O_8$ (x = 0 or 0.5) and $Zr_{0.98}Sc_{0.02}W_2O_{8-y}$, indicators of the order parameter, fitted a universal curve when plotted against the transition temperature. The results suggested that the materials all have the same order-disorder transition mechanism[269]. A sudden decrease in the transition temperatures of $Zr_{1-x}M_xW_2O_{8-y}$ (M = Sc, In, Y), as compared to that of $Zr_{1-x}Hf_xW_2O_8$, was attributed to a decrease in order of the α-phase due to oxygen defects.

An effective way to control the thermal expansion of ZrW_2O_8 is the insertion of NH_3 into the voids of the structure. Studies reveal that the nitrogen bonds to tungsten in the ammoniated material and the latter retains its original $P2_13$ structure[270]. This ammoniation improves the thermal stability, increases the phase transition temperature by about 50K and changes the thermal expansion from -7.8 x 10^{-6} to -2.1 x 10^{-6}/K.

Cubic $ZrWMoO_8$ powders with rod-like aggregate and thin flake-like and flower-like rod cluster morphologies have been made hydrothermally by using various amounts of $(NH_4)_2HPO_4$ as surfactant. The thermal expansion properties were not affected by the addition of $(NH_4)_2HPO_4$. Cubic $ZrWMoO_8$ powders having either rod-like aggregate or flower-like rod cluster morphologies exhibited positive thermal expansion between room temperature and 120C. They exhibited negative thermal expansion between 120C and 700C[271]. The abnormal thermal expansion below 120C was attributed to the presence of water molecules. A precursor dehydration route was found for the synthesis of cubic $ZrW_{1.6}Mo_{0.4}O_8$. The hydrate precursor was dehydrated at 473K and transformed into the cubic compound above 800K. A novel intermediate phase, orthorhombic $ZrW_{1.6}Mo_{0.4}O_8$, occurs between 573 and 800K[272]. Its space group is $Pmn2_1$ and the lattice parameters are: a = 0.5917, b = 0.7273, c = 0.9148nm and Z = 2.

Materials of the form, $ZrW_{2-x}Mo_xO_8$ (0 ≤ x ≤ 2), have been synthesized by reacting a mixture of ammonium tungstate and ammonium molybdate with zirconium oxynitrate using a hydrothermal method. The structural phase transition temperature decreased slightly with increasing substitute content. The cubic-to-trigonal phase transition was also affected by the latter content[273]. The resultant products decomposed into WO_3/MoO_3 and ZrO_2 with increasing temperature when x ≤ 0.5. Cubic solid solutions of the form, $ZrW_{2-x}Mo_xO_8$ (x = 0 to 1.3), were prepared by using a polymorphous precursor transition route[274]. The solid solutions were single-phase with α- and β-ZrW_2O_8 structures for 0 ≤ x ≤ 0.8 and 0.9 ≤ x ≤ 1.3, respectively. Cubic $ZrW_{1.7}Mo_{0.3}O_8$ was synthesized by using zirconium oxynitrate, ammonium tungstate and ammonium molybdate as raw materials[275]. The particle size distribution was relatively uniform. Between room temperature and 700C, the intrinsic and macro thermal expansion coefficients were -6.61 x 10^{-6} and -5.76 x 10^{-6}/K, respectively. Samples of $ZrWMoO_8$ were obtained by using ammonium tungstate, molybdate tungstate and zirconium tungstate as starting materials;

via dehydration of the precursor, $ZrWMoO_7(OH)_2(H_2O)_2$. The type of gelling agent used had a marked effect upon the morphologies of the resultant product[276]. The thermal expansion coefficient of cubic $ZrWMoO_8$, prepared using HCl as the gelling agent, was -3.84 x 10^{-6}/K between 100 and 700C.

Oxygen migration at very low temperatures has been measured in $ZrW_{2-x}Mo_xO_8$ phases. The material exhibits a thermal expansion of -8.7 x 10^{-6}/K and oxygen migration, with an activation energy of 0.23eV, at temperatures as low as 200K. Oxygen migration at such low temperatures is rare in electronic insulators, and is associated here with the order-disorder transition between the α and β forms and a volume change of -0.24%[277]. A negative volume change during an order-disorder transition is also rare.

Molybdenum-substituted thin films, $ZrW_{1.1}Mo_{0.9}O_8$, were deposited onto quartz substrates by means of pulsed laser deposition. The as-deposited films contained amorphous phases, but crystallized cubic $ZrW_{1.1}Mo_{0.9}O_8$ films were obtained by heating (1050C, 420s). The growth of the films was markedly affected by the substrate temperature and the oxygen pressure. Films which were deposited at 500C under an oxygen pressure of 10Pa were smooth and compact, with a thickness of about 720nm. The cubic $ZrW_{1.1}Mo_{0.9}O_8$ exhibited marked negative thermal expansion, with a coefficient of -8.65 x 10^{-6}/K between 100 and 600C. The substitution of molybdenum led to a marked decrease in the α-to-β phase transition to below 100C[278]. With increased temperature, some of the cubic material changed into a trigonal phase. Coatings of $ZrW_{1.5}Mo_{0.5}O_8$ were synthesized by pulsed laser deposition. The thermal expansion coefficient of the sintered (1080C, 6h) material was -7.13 x 10^{-6}/K between 30 and 600C. Smooth compact 835nm coatings were deposited at 500C and annealing (1180C, 420s, air) led to a transition from amorphous to polycrystalline[279]. The average thermal expansion coefficient was -7.7 x 10^{-6}/K between room temperature and 600C. The α-to-β transition temperature again fell to below 100C.

Sintering of TiO_2-WO_3-ZrO_2 mixtures led to the formation of $Zr_{1-x}Ti_xW_2O_8$ solid solutions. A decrease in the unit cell parameter and the order-disorder transition temperature occurred. Previous studies of ZrW_2O_8 solid solutions had attributed an increase in the phase transition temperature to a decrease in the free lattice volume, and a decrease in the phase transition temperature was attributed to the presence of a more disordered state. The present results indicated that the phase transition temperature was markedly affected by the bond dissociation energy of the addition[280]. A decrease in the bond strength could negate the effect of decreased lattice free volume, and lower the phase transition temperature as the degree of Ti^{4+} substitution was increased. When Ti^{4+} and Sn^{4+} were substituted for the Zr^{4+} in ZrW_2O_8, substitution led to a decrease in the cell parameters because the ionic radii of the substituents were smaller than that of Zr^{4+}. A marked decrease in the transition temperature also occurred[281]. The strength of the M-O bond was suggested to play an important role, because it is implicated in both rigid unit mode motion and in the order-disorder transition mechanism.

The solid-solution series, $ZrW_{2-x}V_xO_{8-x/2}$, exhibits thermal expansion properties like those of ZrW_2O_8. At room temperature, cubic $ZrW_{2-x}V_xO_{8-x/2}$ ($x = 0$-0.24), was of α-ZrW_2O_8 type and $ZrW_{1.70}V_{0.30}O_{7.85}$ had β-ZrW_2O_8-type disordering. The average crystal structure of $ZrW_{2-x}V_xO_{8-x/2}$ was a superposition of cubic ZrW_2O_8 and ZrV_2O_7 structures. The lattice parameters of cubic $ZrW_{2-x}V_xO_{8-x/2}$ solid solutions, and the order-disorder transition temperature of the solid solutions, decreased with increasing x[282]. The maximum solid solubility of cubic $ZrW_{2-x}V_xO_{8-x/2}$ solid solution attained 16%.

Solid solutions of the form, $Zr_{1-x}M_xW_2O_{8-y}$ (M = Sc, In, Y) with $x = 0.04$, have been synthesized by solid-state reaction. Between 90 and 560K, all of the solid solutions had a cubic structure and negative thermal expansion coefficients. The lattice parameters of $Zr_{1-x}M_xW_2O_{8-y}$ were smaller than those of ZrW_2O_8 and this was attributed to the effect of oxygen defects; even though the ionic radii of substituted M^{3+} ions are greater than those of Zr^{4+}. The order-disorder transition temperatures of the substituted samples decreased markedly in the order: Y, In, Sc, and decreased with increasing M content[283]. The room-temperature thermal conductivities of ZrW_2O_8 and $Zr_{0.99}Y_{0.01}W_2O_8$ are about 0.80 and 0.65W/mK, respectively; lower than those of other functional oxides. The mechanism of thermal resistance is suggested to be phonon-phonon scattering, and is affected by the phase transition at about 420K[284]. The differing thermal conductivities of $Zr_{0.99}Y_{0.01}W_2O_8$ and ZrW_2O_8 suggest that Y^{3+} acts as a scattering center. The heat capacities of $Zr_{1-x}M_xW_2O_{8-y}$ (M = Sc, Lu; $x = 0.02$, 0.04) at low temperatures were measured. The effective phonon density of states of β-ZrW_2O_8 had a rounded form at about 5meV[285]. This rounded phonon form contrasted with that of the low-temperature phase and the difference was consistent with the differences in negative thermal expansion. The partial substitution of Y^{3+} for Zr^{4+} is effective in promoting grain growth and in increasing the relative density[286]. Negative thermal expansion was observed in all specimens of the form, $Zr_{1-x}Y_xW_2O_{8-\delta}$ ($x = 0.00$ to 0.02), and the coefficient was hardly affected by the Y^{3+} content.

Solid solutions of the form, $Zr_{1-x}Ln_xW_2O_{8-x/2}$ (Ln = Eu, Er, Yb), had coefficients of thermal expansion of the order of -10.3×10^{-6}/K at 30 to 100C[287]. The solubility of the lanthanide ions in the solid solutions decreased linearly with increasing radius of the substituted lanthanide ions.

Composites

One of the many useful applications of negative thermal expansion materials is as composites together with normal materials; thus imparting Invar like properties for example. Finite element analysis of composites involving negative thermal expansion materials and positive thermal expansion materials, such as ZrW_2O_8 in copper or ZrO_2 in ZrW_2O_8, predict that, during rapid temperature changes, the transient short-time thermal expansion can be considerably greater than the steady-state value. Thermal stresses in the composite can be high; particularly at the interfaces between the materials[288]. They can even exceed the material's strength.

Negative thermal expansion ZrW_2O_8 was prepared via step-by-step solid-state reaction of ZrO_2 and WO_3 powders. The coefficient of thermal expansion of the as-prepared ZrW_2O_8 was about -5.08×10^{-6}/K at 20 to 700C. Various amounts of the ZrW_2O_8 powder were then added to polyamic acid so as to form polyimide/ZrW_2O_8 composites. With increasing ZrW_2O_8 content, the powder significantly enhanced the thermal stability of the composites, and reduced the thermal expansion[289]. A 50wt%ZrW_2O_8 addition could lead to a 31% reduction in the coefficient of thermal expansion. When sub-μm particles of ZrW_2O_8 were incorporated into aromatic polyimide, cross-sectional SEM images indicated that the filler particles were homogeneously dispersed; with no aggregation. The coefficients of thermal expansion of films decreased linearly with increasing volume fraction of ZrW_2O_8. The coefficient for a composite film containing 60vol% of the filler was as low as that of metallic copper[290]. Assuming that thermal expansion is three-dimensionally isotropic, composite films containing more than 50vol%ZrW_2O_8 could be expected to exhibit volumetric thermal expansions of less than 50×10^{-6}/K. Nanocomposites with a tunable coefficient of thermal expansion were prepared by incorporating rod-shaped cubic zirconium tungstate nanoparticles into polyimide. The interfacial interaction between the polyimide and the ZrW_2O_8 was increased by covalently bonding various other organics, such as a short aliphatic silane or a polyimide oligomer, to the ZrW_2O_8 surface. The addition of the ZrW_2O_8 nanoparticles did not affect the thermal degradation or glass transition temperatures of the polyimide. The added nanoparticles increased the Young's modulus of the polymer, and the increase was greater for nanocomposites with engineered interfaces; due to the efficient load transfer offered by the linking groups[291]. The coefficient of thermal expansion of the polyimide was reduced by about 22% by adding 15vol% of ZrW_2O_8. A composite was fabricated by allowing the ceramic to settle in epoxy resin before curing, and then using only the dense bottom fraction of the mixture. The samples consisted of some 60vol% of tungstate, with no significant voids. The coefficient of thermal expansion was measured between 25 and 300K, showing that negative thermal expansion occurred below about 100K[292]. This behavior was consistent with the predictions of a variational model which showed that a high ceramic loading is necessary in order to obtain negative thermal expansion of the composite.

A $ZrW_2O_8/Zr_2WP_2O_{12}$ composite was prepared by liquid-phase sintering. The apparent density of the ZrW_2O_8, without any sintering additive was some $3.7g/cm^3$ (about 73% of theoretical density). The relative density of samples which were sintered with more than 5mol% of P_2O_5 was about 90%. The phases present were mainly ZrW_2O_8, with small amounts of WO_3, ZrO_2 and $Zr_2WP_2O_{12}$. It was deduced that the melting of ZrO_2-P_2O_5 resulted in a liquid phase which was then converted into $Zr_2WP_2O_{12}$ in the final stages of sintering. The $Zr_2WP_2O_{12}$ was observed in the gaps between ZrW_2O_8 grains, and at triple junctions[293]. Ceramics which were sintered with 20mol% of P_2O_5 had a thermal expansion coefficient of -4.0 x 10^{-6}/K.

Fully-dense Cu-75vol%ZrW_2O_8 metal composites have been prepared by the hot isostatic pressing of copper-coated ZrW_2O_8 particles. A small amount of high-pressure λ-ZrW_2O_8 was created during cooling and depressurization after densification. Almost complete conversion to λ-ZrW_2O_8 occurred during subsequent cold isostatic pressing. The thermal expansion behavior between 25C and 325C was altered by the cold isostatic pressing treatment. It also depended upon the length of time between thermal cycles. The thermal expansion coefficients ranged from -6 x 10^{-6}/K, to far above the thermal expansion coefficient of the copper matrix[294]. This expansion/contraction behavior was explained in terms of phase transformations occurring within the ZrW_2O_8.

Composites with the composition, (Al_2O_3-20wt%ZrO_2)-ZrW_2O_8, were obtained by cold-pressing and sintering. The composites comprised mainly a monoclinic modification of zirconium dioxide and the orthorhombic phase of aluminum oxide[295]. After adding zirconium tungstate, the phase composition of the sintered ceramics changed; involving the formation of tungsten-aluminate spinels of the form, $Al_x(WO_y)_z$. The thermal expansion coefficient of the composite material decreased by about 30%, as compared with those of the initial components.

The powdered tungstate has been mixed with commercial glue so as to produce extra-low dilation adhesion[296]. Such a composite glue is useful for packaging opto-electronics.

Chapter 6

Halides

It has recently been shown that $CaZrF_6$ combines marked negative thermal expansion, over ranges of 10 to 1000K, with optical transparency between the mid-infrared and the ultra-violet. Such materials are also easily formed ceramics and can be handled in air. In addition, $CaZrF_6$ and $CaHfF_6$ exhibit a much greater negative thermal expansion, -18 x 10^{-6} and -22 x 10^{-6}/K respectively, than do ZrW_2O_8 and other corner-shared framework structures. Their negative expansion is comparable to that reported for framework solids containing multi-atom bridges, such as metal cyanides and metal-organic frameworks. The negative expansion of $CaZrF_6$ is strongly temperature-dependent; first-principles calculations show that it is driven mainly by vibrational modes below about 150/cm. It is elastically soft, with a bulk modulus at 300K of 37GPa. Under compression, it begins to disorder at about 400MPa.

The replacement of calcium by magnesium, and of zirconium by niobium, modifies the behavior of the compound. Like $CaZrF_6$, $CaNbF_6$ retains the cubic ReO_3-type structure down to 10K and exhibits negative thermal expansion up to at least 900K. It undergoes a phase transition under 400MPa compression at room temperature, preceded by marked pressure-induced softening, and exhibits pressure-induced amorphization above 4GPa. The $MgZrF_6$ modification adopts a cubic, Fm3, structure at 300K and undergoes a symmetry-lowering transition involving octahedral tilts at 100K. It exhibits slight negative thermal expansion just above this transition. The thermal expansion increases upon further heating, and passes through zero at 500K. Contrary to $CaZrF_6$ and $CaNbF_6$, it undergoes an octahedral tilting transition at 370MPa during compression before exhibiting a reconstructive transition at 1GPa. The marked negative expansion of $CaZrF_6$, which remains cubic to below 10K, contrasts sharply with cubic $CoZrF_6$, which exhibits only a moderate negative expansion above its rhombohedral-to-cubic transition at about 270K[297]. Cubic $MgZrF_6$ exhibits pressure-induced softening and stiffening during heating. The final modification, $MgNbF_6$, is cubic (Fm3) at room temperature but exhibits a symmetry-lowering octahedral tilting transition at 280K[298]. It does not exhibit any negative thermal expansion between 100 and 950K. It is clearly the replacement of calcium by smaller and more polarizing magnesium which leads to the greater changes in thermal expansion.

ScF$_3$

The cubic ReO$_3$ structural type is often used as a simple example of how negative thermal expansion can arise from the thermally induced rocking of rigid structural units. Scandium fluoride was the first material having this structure which provides clear experimental evidence of that mechanism.

A rapid low-temperature (310C, 0.5h) synthesis route for the preparation of ScF$_3$ is to use NaNO$_3$ or KNO$_3$ as a reaction medium and Sc(NO$_3$)$_3$ and NH$_4$HF$_2$ as precursors The ScF$_3$ thus produced has a relatively regular morphology of high crystallinity or is monocrystalline. The type of molten salt used affects the morphology of the particles: replacing NaNO$_3$ with KNO$_3$, changes cubes to sticks[299]. The molten salt is also important to the elimination of non-stoichiometric ScF$_{2.76}$, a common impurity.

The trifluoride retains its cubic ReO$_3$-type structure down to at least 10K, although the pressure at which the cubic-to-rhombohedral transition occurs falls from 0.5GPa, at about 300K, to between 0.1 and 0.2GPa at 50K. There is a marked negative thermal expansion of -14 x 10^{-6}/K between 60 and 110K[300]. Upon heating, the coefficient of thermal expansion increases smoothly up to a room-temperature value which is similar to that of ZrW$_2$O$_8$. It becomes positive at above about 1100K. Negative thermal expansion in the isostructural cubic ReO$_3$ and ScF$_3$ materials has been studied using first-principles calculations within the quasi-harmonic approximation. The predicted coefficients of negative thermal expansion were in reasonable agreement with experiment. The negative thermal expansion behavior was ascribed to vibrational modes which have minimum negative values of the Grüneisen constants at M(0.5, 0.5, 0) and R(0.5, 0.5, 0.5) in the Brillouin zone. The calculated coefficient for ReO$_3$ is an order of magnitude lower than that for ScF$_3$, due to the stiffer covalent bonds in ReO$_3$ and to the vibrational modes which generate a sharp energy-increasing antibonding state[301]. Its larger lattice constant and smaller bulk modulus, compared to those of ReO$_3$, mean that ScF$_3$ is destined to have a higher coefficient of negative thermal expansion.

The behavior of the material has also been studied by means of density-functional molecular dynamics in the isothermal-isobaric regime, and the simulations reproduce the known trends of negative thermal expansion at low temperatures, its approach to zero at about1000K and its positive nature at higher temperatures. The simulations imply that the observed phenomena arise from the correlated dynamics of ScF$_6$ octahedra[302]. There was also predicted to be a relationship between the cubic-to-rhombohedral transformation under slight compression, and the thermal behavior. Inelastic neutron scattering studies have been made of the temperature dependence of the lattice dynamics from 7 to 750K. The phonon densities-of-state involve a large anharmonic contribution, with thermal stiffening of modes at around 25meV. First-principles phonon calculations identified the individual modes in the densities of states and frozen phonon calculations revealed that some of the modes, with fluorine atom motion transverse to their bond direction, behave as quantum quartic oscillators[303]. The quartic potential arises from harmonic interatomic forces in the DO9 structure, and accounts for phonon stiffening with temperature. It also explains a large part of the negative thermal expansion. *Ab initio* electronic structure and lattice dynamics calculations of cubic and distorted ScF_3 have been performed using the linear combination of atomic orbitals method. The calculated band-gap is 10.54eV[304]. The calculated infra-red spectra of fluorine-atom displaced cubic ScF_3 leads to the prediction that its mean Sc-F-Sc angle during negative thermal expansion deviates from 180°. Other calculations of the temperature-dependent phonon spectra describe quantitatively the behavior of the fluoride and the suppression of negative thermal expansion; leading to the prediction of an anomalous temperature dependence of the thermal conductivity. It was proposed that this is a general feature of the perovskite class[305]. It was noted that the suppression of negative thermal expansion at high temperatures cannot be treated within the quasi-harmonic approximation. Such suppression is better based upon the temperature dependence of the mode Grüneisen parameters.

Diffuse X-ray scattering data show that two-dimensional nanoscale correlations exist in momentum-space regions which are possibly associated with rigid rotations of the perovskite octahedra[306]. A proposed rigid octahedral motion model implies the existence of a complicated link between the nm correlation length scale, the energy scale for octahedral tilt fluctuations and the coefficient of thermal expansion. On the other hand, synchrotron-based X-ray total scattering, extended X-ray absorption fine structure and neutron powder diffraction data suggest that the Sc-F nearest-neighbor distance increases markedly with increasing temperature while the Sc-Sc next-nearest neighbor distance decreases. Meanwhile the thermal ellipsoids of relative vibrations between the Sc-F nearest-neighbors are very elongated perpendicular to the Sc-F bond. The latter observation indicates that the Sc-F bond is much softer in bending than in stretching, rather like a guitar-string, encouraging mainly transverse motion of the fluorine atoms rather than static displacement[307]. It is deduced that rigid unit motion is not implicated in the occurrence of negative thermal expansion; the important factor being the string-like motions. But perovskite-structured materials contain very many ordered phases of electronic or magnetic type, and an always-present influence beyond those interactions is the lattice. There are hints of a quantum phase transition very near to the ground state of the present non-magnetic simple cubic perovskite. It is suggested its properties are strongly affected by these features, up to 100K above the apparent transition[308]. Spatial and temporal correlations in the high-symmetry cubic phase indicate soft-mode, central peak, and thermal expansion phenomena are all strongly affected by these the above factors.

The degree of negative thermal expansion decreases with decreasing crystal size. Specific-heat measurements indicate that the low-energy phonon vibrations, which account for the negative expansion behavior, stiffen as the crystal size decreases. With decreasing size, the peaks in the high-energy X-ray pair distribution function broaden, strongly indicating the occurrence of increased atomic displacements; such as those which are suggested to explain the stiffening of negative-expansion related lattice vibrations[309]. This implies that the negative expansion properties of open-framework materials can be adjusted by controlling the crystal size.

Another routine way of controlling thermal expansion is to make additions to the basic material. Thus $Sc_{1-x}Y_xF_3$ (x ≤ 0.25) was studied using synchrotron powder diffraction between 100 and 800K. Its behavior under pressures of up to 0.276GPa was also determined while heating from 298 to 523K. Just a 5% addition of the larger Y^{3+} ion led to a cubic-to-rhombohedral transition when cooling from room-temperature to 100K. The coefficient of thermal expansion in the rhombohedral phase depended strongly upon both composition and temperature. Above 400K, all of the samples are cubic and the coefficient of thermal expansion is essentially independent of composition. The isothermal bulk modulus and thermal expansion of pure ScF_3, but not those of the present materials, are independent of temperature and pressure, respectively[310]. The yttrium substitution decreases the bulk modulus even when the sample is cubic but all of the solid solutions stiffen upon heating.

Control of the thermal expansion has also been investigated by preparing $Sc_{1-x}Ti_xF_3$ solid solutions; TiF_3 being fully soluble in ScF_3 at 1338K. These solid solutions were characterized by means of synchrotron powder diffraction performed at 100 to 500K. The temperature of the cubic-to-rhombohedral phase transition varied linearly with composition above 100K. At large titanium contents the transition was first-order[311]. The rhombohedral phase of each composition exhibits a very positive thermal expansion whereas the expansion of the cubic phase is negative, between 420 and 500K, for every composition. A first-principles investigation of $Sc_{1-x}Ti_xF_3$ (x ≤ 0.375) predicts that controllable thermal expansion can be achieved by varying the titanium content. The negative thermal expansion is expected to lessen gradually with increasing titanium content[312]. The Jahn-Teller effect is expected to play an important role in the cubic-to-rhombohedral phase transition, since this occurs when increased energy stability results from the 3d orbitals of Ti^{3+} cations splitting into triply-degenerate forms.

TiF$_3$

Among negative thermal expansion materials, this rhombohedral phase is a new member and its properties were predicted on the basis of first-principles calculation. Its negative expansion occurs at low temperatures[313]. The underlying mechanism has been explained in terms of vibrational modes, and the rigid unit mode of internal TiF$_6$ octahedra in the low-frequency optical range is held to be responsible for the negative expansion properties.

ZnF$_2$

This is another material whose negative thermal expansion at lower temperatures has only recently been recognised[314]. First-principles calculations show that ZnF$_2$ is an insulator with a direct band-gap. Strong hybridization occurs between the Zn-3p, 4s and F-2p states. Using the quasi-harmonic approximation, the resultant relationship between volume and temperature confirms the negative-expansion behaviour. All of the phonon vibrational modes contributing to the negative expansion, and identified from the Grüneisen parameters, are low-frequency optical phonons[315]. The lowest-frequency rigid unit mode of ZnF$_6$ causes rotary coupling between two adjacent octahedra and shortens the Zn-Zn separation: thus producing negative expansion in a manner which is ubiquitous, as the many examples given in the present book demonstrate.

Chapter 7

Zeolites

Quantitative variable-temperature neutron powder diffraction studies were made of the pure silica zeolites, chabazite and ITQ-4. Chabazite has a linear expansion coefficient which varies from -0.5 x 10^{-6} to -16.7 x 10^{-6}/K between 293 and 873K[316]. In accord with studies of ZrW_2O_8- and $Sc_2(WO_4)_3$-type materials, the contraction was attributed to changes in the Si-O-Si interpolyhedral bond angles. ITQ-4 also contracts between 95 and 510K, with the coefficient varying from -2.2 x 10^{-6} to -3.7 x 10^{-6}/K. Between 50 and 500C, ITQ-1 has a coefficient of 12.1 x 10^{-6}/K and ITQ-3 has a coefficient of -11.4 x 10^{-6}/K[317]. The zeolite, SSZ-23, has a coefficient of -10.3 x 10^{-6}/K over the same temperature range. Powder X-ray and neutron diffraction studies of calcined siliceous zeolites (ITQ-7, ITQ-9, CIT-5) and aluminophosphates (AlPO4-31, Mg-doped AlPO4-17) were carried out as a function of temperature. Of these materials, ITQ-7, ITQ-9 and MAPO-17 exhibit negative thermal expansion over a wide temperature range. The CIT-5 and AlPO4-31 samples are 'anomalous' in that they expand when heated but, based upon data on seventeen microporous materials, it is clear that negative thermal expansion has to be considered normal here rather than exceptional[318]. Positive expansion is favoured by a structure having a relatively high framework density and a one-dimensional channel system. Siliceous ITQ-29 and dehydrated or hydrated silver zeolite-A have large negative, moderately negative and positive thermal expansion coefficients, respectively[319].

The pair distribution function method was used to obtain some insight into the contraction mechanism of chabazite. By optimizing the structure via free-energy minimization, and by using the reverse Monte Carlo technique, it was possible to find structural models which agreed quantitatively with the experimental pair distribution functions at low and high temperatures. It was concluded from these models that the contraction mechanism could not involve rocking of the tetrahedra as rigid unit modes, because there were distortions of the tetrahedra with temperature[320]. The contraction mechanism was suggested to involve a mode which moved along the Si-O_3-Si-O_4-Si linkages.

Anhydrous AlPO4-17 also exhibits marked negative thermal expansion between 18 and 300K. Powder synchrotron X-ray diffraction data reveal an essentially linear decrease in cell volume, with a linear coefficient of thermal expansion of -11.7×10^{-6}/K. The contraction along the a- and b-axes is considerably greater than that along the c-axis of the hexagonal structure[321]. Rietveld refinement of the framework has indicated that the negative thermal expansion is related to harmonic transverse vibrations of bridging oxygen atoms, leading to rocking of the rigid tetrahedral units of the structure. Free-energy minimization has been applied to hexagonal $AlPO_4$-17 in order to predict its thermal expansion. A marked negative expansion is expected for both the a- and c-axes; as observed in practice. This structure allows for a rocking motion of the rigid polyhedra, but the present calculations of the phonon density-of-states indicated that vibrations of that type are not important for the negative thermal expansion of this compound[322]. The calculations indicated instead that the negative expansion is due to apparent decreases in the bond distances, rather than to a decrease in the Al-O-P angles which was in turn related to transverse thermal motions of the oxygen in the Al-O-P linkages.

High-resolution powder X-ray diffraction studies have been made of structurally-related open-framework phosphates, $AlPO_4$ and $GaPO_4$, having the CHA-type structure (AlPO-34 and GaPO-34, respectively) and $AlPO_4$ having the AEI-type framework (AlPO-18). All three materials exhibit negative thermal expansion. In the case of AlPO-34, this occurs between 110 and 450K. In GaPO-34, the volume expansion is generally negative above 210K while, below, there is a phase transition to a new monoclinic polymorph. In AlPO-18, the behavior of the negative thermal expansion between 110 and 450K can be directly related to AlPO-34. A comparison with data for SiO_2 having the CHA-type structure shows that the iso-electronic $AlPO_4$ has a rather similar thermal behavior (table 4). The magnitude and anisotropy of the thermal expansion of $GaPO_4$ can be tailored[323]. The structural units, double six-rings and eight-ring windows, which are common to each material, behave similarly as a function of temperature.

Orthorhombic hydrated HZSM-5 zeolite, with its silica/alumina molar ratio of 30, exhibits a complex thermal expansion behavior between 40 and 840C. An average overall linear coefficient of thermal expansion of -9.0×10^{-6}/K is found in this region. The expansion is anisotropic, with the a- and c-axes contracting more than does the b-axis. Five thermal expansion regions can be identified: the first region, between 40 and 80C, involves a coefficient of 6.19×10^{-6}/K. The second one, between 80 and 440C, has a coefficient of -7.38×10^{-6}/K. The third region, between 440 and 520C, has a coefficient of 3.89×10^{-6}/K. The fourth has a coefficient of -6.22×10^{-6}/K and the fifth has a coefficient of -25.75×10^{-6}/K. It can be concluded that the release of H_2O molecules does not produce the negative average thermal expansion; from 80 to 440C it can be attributed mainly to the effect of H_3O^+ species. The expansion between 440 and 520C is related to dehydroxylation[324]. The strong negative thermal expansion at above 520C seems to be related to the familiar basic mechanism of transverse vibrations of the bridging oxygen atoms between two rigid polyhedra. Variable-temperature single-crystal X-ray diffraction studies have permitted detailed study of the mechanism of negative thermal expansion in pure silica zeolite, IFR, between 30 and 557K. The structure comprises two regions: columns of fused rings which expand during heating, and inter-column regions which tend to contract during heating[325]. The competing changes lead to a material which contracts parallel to the a- and b-axes but expands along the c-axis.

Table 4 Thermal Expansion Coefficients of Zeolite-Type Materials

Composition	Axis	Temperature Range (K)	Coefficient (/K)
AlPO-34	a	110-460	-9.27×10^{-6}
AlPO-34	b	110-460	-9.27×10^{-6}
AlPO-34	c	110-460	-5.54×10^{-6}
GaPO-34	a	260-460	2.15×10^{-6}
GaPO-34	b	260-460	2.15×10^{-6}
GaPO-34	c	260-460	-20.30×10^{-6}
AlPO-18	a	105-455	-11.94×10^{-6}
AlPO-18	b	105-455	-6.81×10^{-6}
AlPO-18	c	105-455	-7.32×10^{-6}

In situ time-resolved synchrotron X-ray and neutron powder diffraction studies have revealed that the negative thermal expansion and so-called trap-door cation relocations, which are observed in zeolite rho, result from water-mediated chemical changes during dehydration[326].

A marked isotropic negative thermal expansion of -4.2×10^{-6}/K was identified in siliceous faujasite between 25 and 573K and was attributed to transverse vibrations of the bridging oxygen atoms, leading to coupled librations of the SiO_4 tetrahedra [327]. These tetrahedra expand in size with increasing temperature, and also suffer some distortion. These distortions are not however considered to be extensive enough to prevent the SiO_4 tetrahedra from being treated as rigid units[328]. Further research carried out between room temperature and 1123K has shown that the porous phase is very stable when heated, and that no phase transitions or changes in symmetry are observed. The marked negative thermal expansion between 25 and 573K was confirmed, and extended up to about 923K. Above this point, positive thermal expansion was observed. As before, the negative-expansion behaviour was explained in terms of transverse vibrations of the oxygen atoms involved in tetrahedron-O-tetrahedron linkages[329]. Attention was also drawn to other structural features of the faujasite structure; such as distances between adjacent tetrahedral sites, the thickness of the double 6-membered rings and the di-trigonal index of the 6-membered rings.

In situ high-temperature X-ray diffraction studies of monoclinic silicalite-1 (the silica polymorph of ZSM-5) and of the orthorhombic metallosilicate molecular sieve, zirconium silicalite-1 (with Si/Zr = 50), showed that the structure of silicalite-1 collapsed at 1123K to form α-cristobalite. Both materials exhibited a complex thermal expansion behavior between 298 and 1023K. The zirconium silicalite-1 was stable. Powder X-ray diffraction data revealed high negative lattice thermal expansion coefficients, being equal to -6.75 x 10^{-6} and -17.92 x 10^{-6}/K, between 298 and 1023K, for silicalite-1 and zirconium silicalite-1, respectively. The thermal expansion behavior of both materials is anisotropic, with the strength of contraction along the a-axis being higher than that along the b- and c-axes. Three different thermal expansion regions could be identified within the 298 to 1023K range, and these could be correlated with three stages of weight-loss. Between 298 and 423K, the thermal expansions were 2.647 x 10^{-6} and 4.24 x 10^{-6}/K, for silicalite-1 and zirconium silicalite-1, respectively. In the region between 423 and 873K, the coefficients were -7.602 x 10^{-6} and -15.04 x 10^{-6}/K for silicalite-1 and zirconium silicalite-1, respectively. In the region between 873 and 1023K, the coefficients were - 12.08 x 10^{-6} and -45.622 x 10^{-6}/K for silicalite-1 and zirconium silicalite-1, respectively[330]. As usual, the negative expansion was attributed to transverse vibrations of the bridging oxygen atoms, leading to an apparent shortening of the Si-O distances. Further high-temperature X-ray diffraction studies of iron silicalite-1 molecular sieve samples having Si/Fe ratios of 50, 75, 100 or ∞, showed that they all exhibited negative thermal expansion between 373 and 773K. A systematic increase in the negative thermal expansion coefficient occurred with increasing iron content of the framework[331]. The magnitude of the negative thermal expansion increased in the Si/Fe order: ∞ < 100 < 75 < 50. By changing the composition of the framework, an increased negative thermal expansion can thus be obtained.

Chapter 8

Cyanides

As described elsewhere in this volume, the isotropic negative thermal expansion compound, ZrW_2O_8, has been intensively studied since 1996. A variety of later-discovered materials, which exhibit even larger negative thermal expansions than that of ZrW_2O_8, are the CN-bridge compounds which include ferro-electric ceramics and anti-perovskite manganese nitrides. Zinc cyanide, for example, has an isotropic negative thermal expansion coefficient of about -51 x 10^{-6}/K, at 10 to 370K, which is twice as large as that of ZrW_2O_8. The mechanisms which are associated with the negative thermal expansion of these compounds are usually attributed to their vibrational structures or to magnetic and electronic transitions. Interest first arose in connection with Prussian Blue[332]: the thermal expansion behaviors of nanoporous $M^{II}Pt^{IV}(CN)_6 \cdot xH_2O$ ($0 \le x \le 2$; M = Zn, Cd) compounds were investigated by means of single-crystal X-ray diffraction. The dehydrated phases exhibited a negative thermal expansion which was attributed to a population of low-energy transversely vibrating bridging cyanide ligands. Guest molecules in the framework pore structure could dampen the effect of the transverse vibrational modes. The Zn^{II}-containing phase, where the available pore space matched the volume occupied by an individual water molecule, had a much higher coefficient of thermal expansion. The material response changed from positive to negative thermal expansion when the guest-atom was removed. A reciprocal-space dynamic matrix approach indicated the intrinsic geometrical flexibility of framework structures which contain linear M-CN-M linkages. These impart a marked negative thermal expansion. The topologies of simple cyanide-containing framework materials support a huge number of low-energy rigid-unit phonon modes; all giving rise to negative thermal expansion behavior[333]. Experimental data on compounds of the form, $Zn_xCd_{1-x}(CN)_2$, exhibit such behaviour at 25 to 375K.

AgCN, AuCN, CuCN

The lattice parameters were investigated as a function of temperature at 90 to 490K. All of the materials exhibited one-dimensional negative thermal expansion along the -(M-

C≡N)- chain direction[334]. In CuCN, it was -32.1 x 10^{-6}/K, in AgCN it was -23.9 x 10^{-6}/K and in AuCN it was -9.3 x 10^{-6}/K at between 90 and 490K.

$Ag_3[Co(CN)_6]$

Silver hexacyanocobaltate exhibits positive and negative thermal expansions which are an order of magnitude greater than those found in other crystalline materials. The framework expands along one direction at a rate which is comparable to that of the most weakly bound solids. By flexing like chain-link fencing, the framework couples it to contraction along perpendicular directions[335]. This leads to a negative thermal expansion that is 14 times greater than that of the benchmark ZrW_2O_8. Density functional theory calculations provide insights into the origin of these huge thermal expansions. The results confirm that the positive expansion within the trigonal basal plane, and the negative expansion in the orthogonal direction, are coupled via a network of nearly rigid bonds within the chains of Co-C-N-Ag-N-C-Co linkages. The magnitudes of the coefficients of thermal expansion arise from a very shallow energy-surface. The latter permits flexing of the structure at a low energy cost. The thermal expansion is possible with a moderate value of the overall Grüneisen constant[336]. The above energy-surface is so shallow that it is necessary to incorporate a small empirical dissipative interaction in order to furnish ground-state lattice parameters which match the experimental data at low temperature. A systematic study of $M[AuX_2(CN)_2]$-type materials revealed a negative thermal expansion component along the cyano-bridged chains, with coefficients of -13.7 x 10^{-6}, -14.3 x 10^{-6} and -11.36 x 10^{-6}/K for $Ag[AuCl_2(CN)_2]$, $Ag[AuBr_2(CN)_2]$ and $Cu[AuBr_2(CN)_2]$, respectively. The Au•••X and Ag•••X interactions affected the thermal expansion in a similar manner to that of metallophilic Au•••Au interactions in $M[Au(CN)_2]$ and AuCN. Replacing the chlorine with larger bromine atoms had a less significant effect[337]. A similar study of the $M[Au(CN)_4]$ series, where the volume thermal expansion coefficient is 41 x 10^{-6} and 68.7 x 10^{-6}/K for M = Ag and Cu, respectively, confirmed the importance of the effect of atomic radius on the flexibility of the framework and thus upon the thermal expansion properties.

$CdPt(CN)_6$, $NiPt(CN)_6$

The Prussian Blue analogues, $M^{II}Pt^{IV}(CN)_6$ (M^{II} = Cd, Co, Cu, Fe, Mn, Ni, Zn), were investigated using powder X-ray diffraction methods at 100 to 400K. The negative thermal expansion varied widely, depending upon M^{II}, from -1.02 x 10^{-6}/K for $NiPt(CN)_6$ up to a maximum of -10.02 x 10^{-6}/K for $CdPt(CN)_6$. The trend in the magnitude of the negative expansion behavior was in the order: Mn, Fe, Co, Ni, Cu, Zn, Cd; paralleling the order of cation size[338]. Analysis of the temperature-dependence of the average structures suggested that differences in the thermal expansion were due mainly to the differing strengths of the metal-cyanide binding. The plain $Cd(CN)_2$ itself exhibits a negative expansion of -33.5 x 10^{-6}/K between 170 and 375K[339].

In[Ag(CN)$_2$]$_3$

Isostructural di-cyanometallate coordination compounds, $In[M(CN)_2]_3$ (M = Ag, Au), $KCd[M(CN)_2]_3$ and $KNi[Au(CN)_2]_3$, were synthesized and shown to possess a trigonal unit cell. They exhibited positive thermal expansion in the ab-plane, with its Kagome sheets of M-atoms, and negative thermal expansion along the trigonal c-axis; perpendicular to the sheets. The size of the thermal expansion was always unusually large (40 x 10^{-6} to 110 x 10^{-6}/K). The system having the weakest metallophilic interactions, $In[Ag(CN)_2]_3$, exhibited the highest thermal expansion: 105 x 10^{-6}/K along the a-axis and -84 x 10^{-6}/K along the c-axis at 295K. Systems with stronger Au-Au interactions exhibited a relatively small thermal expansion[340]. It was deduced that strong metallophilic interactions hinder huge thermal expansions. The presence of K^+ counter-ions also reduces the degree of thermal expansion.

Ni$_2$[W(CN)$_8$]

The three-dimensional $[Ni^{II}(H_2O)_2]_2[W^{IV}(CN)_8]\cdot4H_2O$ framework undergoes monocrystal-to-monocrystal transformation, during dehydration, to give anhydrous $Ni_2[W(CN)_8]$. While the hydrate exhibits the usual positive thermal expansion, the anhydrous form is the first known octacyanido-based solid to exhibit negative thermal expansion[341]. The marked change in physical properties is attributed to the removal of the water molecules; thus leaving behind a cyanido-bridged bimetallic network accompanied by transformation of octahedral paramagnetic $[Ni^{II}(H_2O)_2(NC)_4]_2$ to square-planar diamagnetic $[Ni^{II}(NC)_4]_2$ components.

Si(NCN)$_2$

Ab initio molecular dynamics simulations and high-temperature synchrotron crystallography reveal that β-$Si(NCN)_2$ exhibits isotropic negative thermal expansion between 500 and 1123K and over its entire thermal stability range. The linear expansion coefficient changes from -1.24 x 10^{-6} at 500K to -1.93 x 10^{-6}/K at 1123K. Study of the vibrational motions traces the effect to transverse vibrations of carbodiimide fragments which bridge Si atoms: Si-N=C=N-Si[342]. Detailed atomistic analysis models the rotational-vibrational movements of the bridging NCN units as a centrifugal governor pulling the crystal framework together as the temperature increases.

Zn(CN)₂

This is the most widely studied member of the class. The structure of this cyanide-bridged material has been probed by means of an atomic pair distribution function analysis of high-energy X-ray scattering data obtained at 100 to 400K. The temperature-dependence of atomic separations extracted from the pair distribution functions indicated an increase in the average transverse displacement, with increasing temperature, of the cyanide bridge from a line connecting the Zn^{II} centers. This then permitted a contraction of the non nearest-neighbor Zn-Zn' and Zn-C/N distances, in spite of the expansion of the individual direct Zn-C/N and C-N bonds. This confirmed that an increase in the average displacement of the bridging atoms is responsible for the negative thermal expansion. The lattice parameters reflected a slight lessening of the negative behavior at high temperatures, as compared with the minimum coefficient of thermal expansion (-19.8 x 10^{-6}/K) which occurred below 180K[343]. This was explained in terms of interactions between the interpenetrating frameworks which make up the overall structure. Density functional theory calculations show meanwhile that the transverse acoustic modes give rise to quite a sharp characteristic, at 2 to 4meV, in the phonon density-of-states[344]. Calculation of the Grüneisen constant of these modes shows that they have large negative values; thus indicating that they are primary causes of the negative thermal expansion. High-pressure Raman spectroscopy of the soft phonons showed that, among the 11 zone-center optical modes, 6 modes had a negative Grüneisen constant. Calculations suggested that the soft phonons corresponded to librational and translational modes of the CN rigid unit, with the librational modes making a greater contribution to the thermal expansion[345]. Rapid disordering of the lattice was detected at above 1.6GPa. First-principles plane-wave pseudopotential studies, based upon density functional theory, of the elastic constants, phonon modes and Grüneisen parameters of $M(CN)_2$-type materials predict that MC_2N_2-MC_2N_2 is the most favorable configuration for $Cd(CN)_2$ while all three possible configurations are almost equally favoured for $Zn(CN)_2$. The $M(CN)_2$ framework includes much stiffer and weaker M-C/N bonds; thus accounting for flexing of the M-CN-M linkage during transverse motion of the cyanide bridge. Transverse vibrations of the carbon and nitrogen atoms in the same (transverse translational mode) or opposite (librational mode) direction both pull the metal atoms closer[346]. The lowest-energy transverse translational optical, modes which are not Raman or infrared active, cause the largest contribution to negative thermal expansion. The pressure dependence of the phonon spectra, at up to 30meV, of polycrystalline samples of $Zn(CN)_2$ under 0, 0.3, 1.9 or 2.8kbar at 165 and 225K indicate the energy-dependence of the ratio of the Grüneisen constant - as a function of phonon energy at ambient pressure – to the bulk modulus[347]. This in turn reflects the phonon anharmonicity. It is concluded that the phonon modes with energies below 15meV play an important role in the negative thermal expansion behavior.

Study of the effect of pressure upon the thermal expansion, using *in situ* neutron powder diffraction experiments at up to 0.6GPa and a third-order Birch-Murnaghan equation to fit the data, showed that the $Zn(CN)_2$ framework became anomalously more compressible at higher pressures. At 50 to 300K, under 0.2 or 0.4GPa, the negative thermal expansion effect became greater with increasing pressure; the coefficient of thermal expansion varying by about -1 x 10^{-6}/K per 0.2GPa of applied pressure up to an average value, for this temperature range, of -19.42 x 10^{-6}/K at 0.4GPa[348]. This behaviour was attributed to an increased framework flexibility at high pressures. Molecular dynamics simulations of such anomalous behavior link the properties in energy space to those in real space and offer some insight into the odd properties[349].

Solid-state ^{67}Zn nuclear magnetic resonance studies confirm that there is head-to-tail disorder of the C-N groups in the solid, and provide information on the relative abundances of the various $Zn(CN)_{4-n}(NC)_n$ tetrahedral species. The latter do not obey a simple binomial distribution: $Zn(CN)_4$ and $Zn(NC)_4$ occur with much lower probabilities than are predicted by binomial theory. This suggests that they have a higher energy than that of the other local arrangements. The lowest-energy arrangement is $Zn(CN)_2(NC)_2$. Total neutron diffraction studies at 11.4K show that the Zn-N and Zn-C bond lengths are 1.969 and 2.030Å, respectively. The main motions which give rise to negative thermal expansion are those in which the carbon and nitrogen atoms within individual Zn-C-N-Zn linkages are displaced to the same side of the Zn---Zn axis[350]. Displacements of the carbon and nitrogen atoms to opposite sides of that axis make negligible contributions at up to 295K.

Single-crystal X-ray diffraction studies at 150 to 300K have provided strong experimental evidence for the existence of low-energy vibrational modes that involve the off-centering of Cd^{2+} ions. These modes effectively increase the network packing density; thus suggesting a mechanism for the negative thermal expansion. Strong local correlations in the displacement directions of neighboring cadmium centers have been detected[351]. Monte Carlo simulations suggest that the results can be interpreted in terms of so-called ice-rules which draw an analogy between the dynamics of $Cd(CN)_2$ and the proton ordering in cubic ice-VII.

$Zn_3[Fe(CN)_6]_2$

This cubic Prussian Blue analogue has been studied using X-ray powder diffraction and inelastic neutron scattering techniques. The former data revealed negative thermal expansion at 300 and 84K, and the associated coefficient was -31.1 x 10^{-6}/K. The neutron vibrational spectrum for $Zn_3[Fe(CN)_6]_2 \cdot xH_2O$, was studied in detail. The inelastic neutron scattering spectrum revealed well-defined and well-separated bands which corresponded to the stretching and deformation modes of the Fe and Zn octahedra[352]. They were all below 800/cm.

Chapter 9

Carbon

A special system of carbon-fibre composites and metals, having an extremely negative value of the thermal expansion coefficient, has been developed. The value is about three times that of steel; with a negative sign. Other advantages were a very low thermal conductivity and a high compressive strength.

C_{60}

A model of an atomic impurity in an octahedral void of C_{60} has been proposed, and solved in the spherical oscillator approximation[353]. It was concluded that such an impurity could contribute to a negative thermal expansion at low temperatures and produce a Schottky-like maximum at higher temperatures. The high-resolution far-infrared vibrational properties of C_{60} fullerene, determined as a function of temperature, reveal an anomalous softening of the $F_{1u}(1)$ mode (526/cm) from 300 to 10K; suggesting that the fullerene cage may expand at low temperatures[354]. These results were consistent with a molecular dynamics study which predicted the occurrence of negative thermal expansion in carbon fullerenes. The thermal expansion of C_{60} has been measured between 2 and 9K using compacted samples with a diameter of about 6mm and height of 2.4mm[355]. The linear coefficient of thermal expansion became negative below about 3.4K.

LaC_2

The coefficients of thermal expansion, and the negative thermal expansion mechanism of tetragonal LaC_2, have been examined using density functional theory and the quasi-harmonic approximation, Numerical results showed that there is an obvious negative thermal expansion parallel to the c-axis, with a coefficient of about -1.67×10^{-6}/K; agreeing well with the measured value of -1.0×10^{-6}/K[356]. A very small negative-expansion along the a-axis was predicted to occur below 10K.

$Yb_{2.75}C_{60}$

Powder synchrotron X-ray diffraction studies of the mixed-valence rare-earth fulleride at 5 to 295K revealed a negative thermal expansion upon cooling below 60K, and indicated the occurrence of a temperature-induced valence transition of the ytterbium atoms. The transformation was electronic, and was driven by a coupling of the ytterbium 4f band and a C_{60} band[357]. It was analogous to one which is observed at lower temperatures in $Sm_{2.75}C_{60}$. It was absent when the electronically active 4f sub-lattice was missing in the related alkaline earth fulleride, $Ca_{2.75}C_{60}$.

Graphene

Negative thermal expansion has been observed in graphene oxide. Between room temperature and 150C, the paper-like graphene oxide sheet exhibited a constant thermal expansion coefficient of -67 x 10^{-6}/K along the in-plane direction[358]. Hysteresis loops were observed in the thermal expansion versus temperature curves, and were affected by the cooling-rate and the starting temperature of the cooling. The intrinsic ripples of suspended graphene can be reversibly controlled by exploiting its negative thermal expansion. Lattice dynamics analysis suggests that, at low temperatures, both phenomena are intrinsic to any two-dimensional crystal with a honeycomb structure or to any monatomic two-dimensional crystal[359]. The negative thermal expansion is caused mainly by the so-called vibrational elongation effect due to large out-of-plane fluctuations, according to calculations based upon self-consistent phonon theory. The thermal expansion coefficient of monolayer graphene was estimated by using Raman spectroscopy at 200 to 400K, and was found to depend strongly upon temperature while remaining negative over the entire temperature range, with a room temperature value of -8.0 x 10^{-6}/K[360]. The strain caused by expansion mismatch between the graphene and the substrate however plays an important role in determining the apparent physical properties of the graphene. The effect of boron and nitrogen doping upon the coefficient of linear thermal expansion was calculated. The coefficient for graphene itself was negative between 0 and 1000K, and its room-temperature value was about -3.26 x 10^{-6}/K. This value became more negative upon B/N doping. To elucidate the cause of the negative thermal expansion, calculations were made of the contributions of individual phonon modes of vibration[361]. It is mainly the transverse acoustic modes which cause the negative expansion.

Chapter 10

Manganese Nitrides

In recent years, anti-perovskite manganese nitrides - which have a similar crystal structure to that of perovskite manganese oxides - have attracted great attention because of their giant and isotropic negative thermal expansion together with magnetostriction, magnetovolume effect, thermal and electrical conduction and nearly zero temperature coefficient of resistance[362]. First-principles calculations of the negative thermal expansion properties of $Mn_3(A_{0.5}B_{0.5})N$ compounds, with A = Cu, Zn or Ag and B = Si, Ge or Sn, predict that - except for $Mn_3(Cu_{0.5}Si_{0.5})N$ - all of these doped compounds will exhibit negative thermal expansion. In the case of the dopants on the B sites, the working temperature of negative thermal expansion shifts to higher temperatures in going from silicon to tin. Among the compounds containing these dopants, $Mn_3(A_{0.5}Ge_{0.5})N$ has the highest negative thermal expansion coefficient[363]. With regard to the dopants at the A sites, when compared to $Mn_3(Cu_{0.5}B_{0.5})N$, $Mn_3(A_{0.5}B_{0.5})N$ (A = Ag or Cd) exhibits negative thermal expansion with higher temperature ranges and lower coefficients of thermal expansion. A particularly important application of such nitrides is as additives which will reduce the thermal expansion of another material. Plastics for example offer low weight and high workability, but their high thermal expansion can be a shortcoming. The large negative thermal expansions of the manganese nitrides can limit the large positive expansions of plastics by forming them into composites with the latter. Fine-grained manganese nitrides have been investigated for the thermal expansion control of light-transmitting plastics[364]. In the grain-refining process, the negative expansion of manganese nitrides changed because of a change in the chemical composition. It is thus necessary to control the negative expansion and grain size simultaneously. Polymer-matrix composites have been filled with high negative thermal expansion materials by means of injection moulding. The huge negative expansion of the nitride can compensate for the large positive thermal expansion of polyamide-imide polymers, and vary the expansion from positive to negative[365]. Quantitative analysis reveals that the thermal expansion of the composite is less than estimates based upon the volume-weighted sum. Copper-matrix composites containing antiperovskite manganese nitrides have been formed by using pulsed electric current sintering, in which the chemically reacted region extended over 10μm around the matrix/filler interfaces. The small-dimension filler deteriorated chemically during the formation of the composite and it lost its negative-expansion capability. It was therefore necessary to produce composites by using only particles having diameters greater than $50\mu m$[366]. The larger filler then suppressed the thermal expansion of copper and improved the composite conductivity to match that of pure aluminum. Looking now in more detail at some specific nitrides:

$Mn_3(Ag,Mn)N$

Study of antiperovskite compounds of the form, $Mn_3Ag_xMn_yN$, with silver vacancies and manganese doping at the silver site showed that the introduction of the silver vacancies had very little influence on the magnetic transitions[367]. The magnetic transitions at the Néel temperature and at lower temperatures gradually overlapped however with increased manganese doping and were associated with broadening of the negative thermal expansion behavior.

$Mn_3(Ag,Zn)N$

The thermal expansion of Mn_3ZnN which had been doped with Ag and sintered at 1223K was measured. The compound has a cubic antiperovskite Mn_3CuN-type structure and exhibits a thermal expansion of -81.00×10^{-6}/K in the case of $Mn_3Zn_{0.5}Ag_{0.5}N$, with an operating width of 18K[368].

$Mn_3(Cu,Ga)N$

With increasing manganese doping of cubic $Mn_3(Ga_{0.7}Cu_{0.3})_{1-x}Mn_xN$ ($x \leq 0.4$) compounds, the temperature range negative-expansion behaviour was shifted to lower temperatures and broadened due to increasing ferromagnetism. A coefficient of linear thermal expansion of -22.8×10^{-6}/K) was found for $x = 0.25$ and 0.3 at temperatures below 120K[369]. Upon adding the $x = 0.3$ powder to epoxy resin, the thermal expansion of the composite was effectively reduced. Almost zero thermal expansion (about 1×10^{-6}/K) and markedly better thermal conductivity were obtained at cryogenic temperatures for composites containing 50vol% of the present filler.

$Mn_3(Cu,Ge)N$

Germanium is regarded as having somewhat of a magical effect upon the antiperovskite manganese nitrides, largely with regard to the marked volume change which is associated with magnetic ordering and its broadening by substitution[370]. There is evidence for a local lattice distortion, described by a low-temperature tetragonal Mn_3GeN structure, while the overall structure remains cubic[371]. This structural instability is closely related to the growth of the ordered magnetic moment; thus triggering a broadening of the volume change. In spite of this enthusiasm for germanium, such antiperovskites have been prepared without using expensive germanium. It is found that tin broadens the volume change, though less effectively than germanium[372]. Substitution of tin widens the operating range of negative expansion almost as much as does germanium.

The negative thermal expansion of antiperovskite manganese nitrides having micron-scale, sub-micron scale or nanometer-scale microstructures were investigated by using $Mn_3Cu_{0.5}Ge_{0.5}N$ as an example. The negative expansion start temperature, operating temperature range and coefficient changed considerably upon decreasing the grain size of the microstructure[373]. The widening of the operating temperature range and a decrease in the absolute value of the expansion coefficient were attributed to a grain-size dependence of the frustrated magnetic interactions and magnetic ordering. When crystallites of $Mn_3Cu_{0.6}Ge_{0.4}N$ of various sizes were prepared by using spark plasma sintering[374], the negative thermal expansion range for 12nm crystallites was 140K; about 75% wider than that for 74nm crystallites. As compared with coarse-grained bulk compound, the negative-expansion start temperature of ultrafine bulk $Mn_3(Cu_{1-x}Ge_x)N$, with x = 0.5, decreased to 250K and the coefficient of thermal expansion reached -23.7 x $10^{-6}/K^{375}$. The nitrogen content of the initial nitride also had a marked effect upon the negative expansion start temperature, operating range and coefficient of thermal expansion. The giant negative thermal expansion found in anti-perovskite manganese nitride can be controlled by the contents of doped germanium and nitrogen vacancies[376]. Calculations indicate that germanium atoms increase the negative thermal expansion, while existing nitrogen vacancies have a weak effect upon the expansion.

The giant negative thermal expansion of the Ge-doped antiperovskite, Mn_3CuN, was studied theoretically by using first-principles calculations. The high negative thermal expansion was attributed to the magnetic phase transition, rather than to lattice vibrations of the Ge-doped compound[377]. The doped germanium atoms significantly increased the antiferromagnetic coupling between nearest-neighbor manganese ions; effectively stabilizing the magnetic ground states. The antiperovskite $Mn_3Cu_{1-x}Ge_xN$, with x ~ 0.5, has a giant negative thermal expansion coefficient; due to the magnetovolume effect near to room temperature. Competition between the $\Gamma 5g$ and $\Gamma 4g$ antiferromagnetic structures around the magnetic ordering temperature was suggested to be the driving force[378]. It was concluded that, in $Mn_3Cu_{1-x}Ge_xN$ (x ~ 0.5), the $\Gamma 4g$ antiferromagnetic component is nearly absent and that competition between the $\Gamma 5g$ and $\Gamma 4g$ antiferromagnetic structures is irrelevant to the giant negative thermal expansion coefficient. The $\Gamma 5g$ antiferromagnetic ordered moment develops gradually with decreasing temperature. The magnetovolume effect is induced by a small amount of germanium at the copper site, and this seems to coincide with recovery of the cubic structure[379]. With further germanium doping, the volume expansion becomes gradual and a room-temperature thermal expansion coefficient of -12 x 10^{-6}/K is found for $Mn_3(Cu_{0.5}Ge_{0.5})N$[380].

The basic mechanical properties of $Mn_3Cu_{1-x}Ge_xN$ were determined, starting with Mn_3CuN: a soft and ductile intermetallic compound with a linear thermal expansion coefficient of 20 x 10^{-6}/K at room temperature. The thermal expansion properties were markedly changed by germanium doping, leading to a coefficient of -12.5 x 10^{-6}/K upon replacing half of the copper with germanium. The associated mechanical properties were also strongly affected by germanium doping, with both the stiffness and hardness increasing with increasing x[381]. Thus $Mn_3Cu_{0.5}Ge_{0.5}N$ had a hardness of 570 to 620HV and a Young's modulus of 211 to 320GPa.

When $Mn_3(Cu_{0.53}Ge_{0.47})N$ was prepared by solid-state sintering in nitrogen, the coefficient of thermal expansion between 200 and 233K was up to -93 x 10^{-6}/K[382]. The negative thermal expansion was attributed to a magnetovolume effect, when the antiferromagnetic spin structure gradually changed to a paramagnetic one. The linear thermal expansion of $Mn_3(Cu_{1-x}Ge_x)N$ ($0.40 \leq x \leq 0.60$), prepared by solid-state sintering in pure nitrogen at 1073K, was negative near to the Neel temperature. With increasing germanium content, the temperature at which negative thermal expansion occurred increased while the temperature range widened and the thermal expansion coefficient decreased. Samples with $x = 0.60$ exhibited linear expansion coefficients of up to -65 x 10^{-6}/K between 250 and 290K[383]. The behaviour was again attributed to a magnetovolume effect. In $Mn_3(Cu_{0.6}Si_xGe_{0.4-x})N$ ($x = 0$ to 0.2), the temperature of transition to negative expansion decreased and the operating range broadened with increasing silicon content[384]. The coefficient of thermal expansion of $Mn_3(Cu_{0.6}Si_{0.15}Ge_{0.25})N$ could attain -16 x 10^{-6}/K, between 80 and 300K, and the operating range could be as wide as 100K. When bulk samples of $Mn_3(Cu_{0.6}Nb_xGe_{0.4-x})N$ (with $x = 0.05$, 0.1, 0.15, 0.2 or 0.25), $Mn_3(Cu_{0.6}Ge_{0.4})N$ and $Mn_3(Cu_{0.7}Ge_{0.3})N$ were prepared by mechanical ball milling and solid-state sintering, the temperature range of negative expansion was about 95K for $Mn_3(Cu_{0.6}Nb_{0.15}Ge_{0.25})N$ and $Mn_3(Cu_{0.6}Nb_{0.2}Ge_{0.2})N$; twice as large as that for $Mn_3(Cu_{0.7}Ge_{0.3})N$[385]. The thermal expansion of $Mn_3(Cu_{0.6}Nb_{0.15}Ge_{0.25})N$ could attain -19.5 x 10^{-6}/K between 165 and 210K. The electrical conductivity was of the order of 2.5 x 10^6S/m. Thermal expansion coefficients of up to -25 x 10^{-6}/K have been found in Ge-doped Mn_3AN, where A is copper, zinc or gallium[386]. The discontinuous lattice expansion observed in pure Mn_3AN is broadened by germanium up to an operating range of 100K at room temperature.

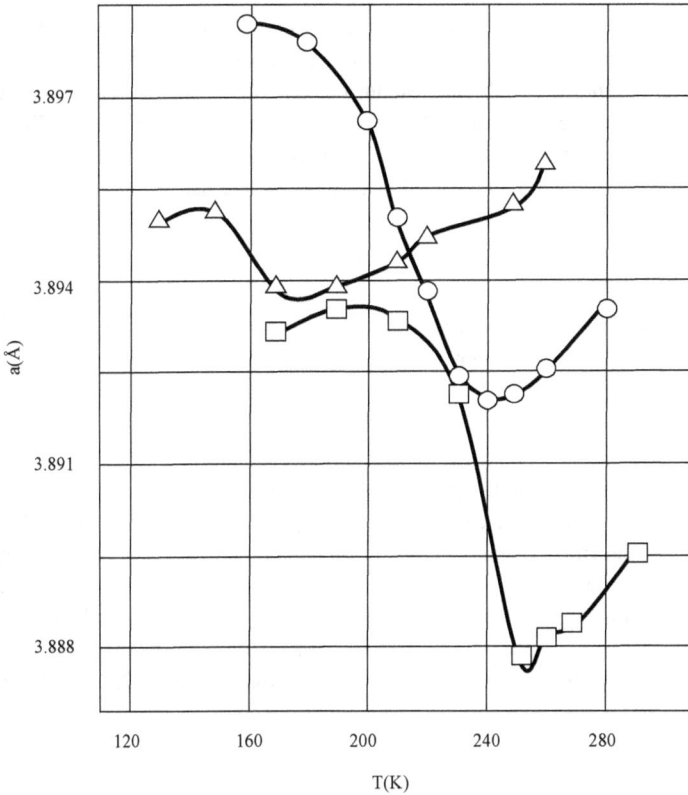

Figure 4 Lattice constant of $Mn_3Ni_{1-x}Cu_xN$ as a function of temperature. Circles: x = 0.5, triangles: x = 0.7, squares: x = 0.3

In $Mn_3Cu_{0.7}Ge_{0.3}N$, the temperature dependence of the interatomic distance reflected the macroscopic negative thermal expansion for both the Cu-Mn and Ge-Mn shells; although the magnitude of the relative change was much greater for the latter than for the former. An enhanced anomaly in the Debye-Waller factor was observed in the Ge-Mn shell within the temperature region of the negative expansion; indicating the presence of static local disorder around the added germanium[387]. The local structural anomalies suggested that local inhomogeneous strain around the germanium was essential to broadening of the discontinuous volume contraction. As a practical test, epoxy composites filled with nanoparticles of $Mn_3(Cu_{0.6}Si_{0.15}Ge_{0.25})N$ modified by plasma treatment were prepared and tested at 77 to 300K. When compared to the bare epoxy resin, the composites had markedly lower coefficients of thermal expansions and higher thermal conductivities. In the case of a composite containing 32vol%nitride, a reduction in the coefficient of thermal expansion of up to 42% was observed between 77 and 195K[388]. The thermal conductivities were 2.8 and 4 times as high as those of the bare resin at 298 and 77K, respectively.

With regard to other properties, $Mn_3(Cu_{0.6}Si_xGe_{0.4-x})N$ (x = 0.05, 0.1, 0.15) - prepared by reactive sintering under pressure - had electrical resistivities and thermal conductivities which ranged from 2.5 x 10^{-6} to 4.3 x $10^{-6}\Omega m$ and 1.9 to 3.6W/mK, respectively[389]. The compressive strength and Young's modulus was about 700MPa and 110GPa, respectively.

$Mn_3(Cu,Ni)N$

The effect of copper doping on the properties of $Mn_3Ni_{1-x}Cu_xN$ (x = 0, 0.3, 0.5, 0.7, 1.0) was that copper at the nickel site complicated the magnetic ground states; thus inducing a competition between antiferromagnetic and ferromagnetic interactions. Negative thermal expansion due to the magnetic ordering transition was observed (figure 5)[390]. For $Mn_3Ni_{0.7}Cu_{0.3}N$, $Mn_3Ni_{0.5}Cu_{0.5}N$ and $Mn_3Ni_{0.3}Cu_{0.7}N$, the ranges within which negative-expansion effects were observed were 190 to 250K, 160 to 240K and 150 to 170K, respectively, while the linear coefficients of thermal expansion were 23.2 x 10^{-6}, 22.3 x 10^{-6} and 15.5 x 10^{-6}/K, respectively.

$Mn_3(Cu,Sn)N$

Upon substituting tin for copper in $Mn_3Cu_{0.8-x}Sn_xMn_{0.2}N$, the original cubic-to-tetragonal phase transition disappears for x ≥ 0.10 and is replaced by a discontinuous lattice expansion with a cubic structure[391]. This discontinuous lattice expansion broadens with increasing dopant level and thermal expansion coefficients of up to -64.54 x 10^{-6}/K are found between 190 and 235K when x = 0.3. Antiperovskite $Mn_3Cu_{0.6}Ag_xSn_{1-x}N$ (x = 0 to 0.3) samples were prepared by mechanical ball-milling and solid-state sintering, and the negative-expansion behaviour was investigated at 80 to 300K. The transition temperature for negative expansion gradually moved towards lower temperatures and the operating range became narrower with increasing silver content[392]. The electrical resistivity exhibited metallic behavior, while the variation in electrical resistivity was almost independent of temperature at above the transition temperature for negative-expansion behaviour.

Mn$_3$GaN

Anti-perovskite manganese nitrides of the form, $Ga_{1-x}N_{0.8}Mn_{3+x}$ ($0 \leq x \leq 3$), were prepared by solid-state reaction. With increasing Mn-doping, the temperature span of the magnetovolume effect broadened and shifted to lower temperatures. For samples with x = 0.2 or 0.25, the temperature span of the negative expansion was 294 to 339K and 255 to 309K, respectively. The corresponding coefficient of linear thermal expansion was -51 x 10^{-6} and -42 x 10^{-6}/K, respectively[393]. An abnormal behavior of the resistivity paralleled the contraction of the lattice. The negative expansion behavior was attributed to a disturbance of the Γ5g antiferromagnetic ground state, caused by partially replacing the gallium atoms with manganese. When nanocrystalline GaN_xMn_3 powders were prepared by mechanical milling, the micrograin material exhibited a sudden volume contraction at the antiferromagnetic-to-paramagnetic transition. The temperature range of volume contraction broadened to 50K as the average grain size decreased to 30nm. The corresponding coefficient of linear thermal expansion attained -70 x 10^{-6}/K[394]. Further reducing the grain size to 10nm caused the temperature range to exceed 100K while the coefficient of thermal expansion remained at up to -30 x 10^{-6}/K for x = 1.0 and at up to -21 x 10^{-6}/K for x = 0.9.

Mn$_3$(Ga,Mn)N

Negative thermal expansion was investigated in $Mn_3Ga_{1-x}Mn_xN_{0.8}$ ($0.1 \leq x \leq 0.3$). As x increased, the temperature range within which the lattice contracted upon heating broadened and shifted to lower temperatures. A coefficient of linear thermal expansion of more than -40 x 10^{-6}/K, with a temperature interval of about 50K, was found around room temperature for x = 0.2 and 0.25. A local lattice distortion which was suggested to be closely related to the negative expansion was invisible in the X-ray pair distribution function of x = 0.3. Zero-field cooling exchange bias was observed as a result of competing ferromagnetic and antiferromagnetic order. The associated ferromagnetic order was an impediment to the growth of antiferromagnetic order and thus broadened the temperature range of negative thermal expansion[395]. The results implied that negative expansion can be achieved in antiperovskite manganese nitrides by manipulating magnetic ordering, without distorting the local structure.

Mn$_3$(Ga,Si)N

A study of $Mn_3Ga_xSi_{1-x}N$ showed that, for pure Mn$_3$GaN, there was a large negative-expansion effect which corresponded to an antiferromagnetic to paramagnetic transition. When the gallium was partially replaced by silicon, the negative-expansion behaviour around the magnetic transition changed. The negative-expansion temperature range broadened to 148K for $Mn_3Ga_{0.75}Si_{0.25}N$ and the linear thermal expansion coefficient was -14 x 10^{-6}/K between 272 and 420K[396]. Upon increasing the silicon content to x = 0.5, the magnetic transition still occurred but there was no negative-expansion effect.

$Mn_3(Ga,Sn)N$

Antiperovskite nitrides of the form, $Mn_3Ga_{1-x}Sn_xN$ (x = 0, 0.1, 0.2, 0.3, 0.4), were prepared by mechanical ball-milling followed by spark plasma sintering. There was a negative-expansion operating range which shifted to higher temperature with increasing tin concentration. The coefficient of linear thermal expansion of $Mn_3Ga_{0.9}Sn_{0.1}N$ typically attained -27.5 x 10^{-6}/K, with an operating range of 59K between 279 and 338K. The coefficient of thermal expansion of $Mn_3Ga_{0.6}Sn_{0.4}N$ was very low between 363 and 400K[397]. The thermal conductivity was about 3.2W/mK at around room temperature, and the compressive strength was about 210MPa.

$Mn_3(Ge,Zn)N$

Antiperovskite $Mn_3Zn_{0.6}Si_xGe_{0.4-x}N$ (x = 0, 0.1, 0.15, 0.2) $Pm\bar{3}m$-structured materials were prepared by spark plasma sintering. The starting temperature for negative-expansion behaviour shifted towards lower temperature, while the operating range of that behaviour was maintained, with increasing silicon doping. The paramagnetic-to-antiferromagnetic transition at about 370K disappeared upon silicon doping[398]. This demonstrated the close relationship between negative thermal expansion and the magnetic transition. When samples of $Mn_3Zn_{0.6}Ge_{0.4}N$ having various crystallite sizes were prepared by spark plasma sintering at differing sintering temperatures, the width of the negative-expansion operating range broadened with decreasing crystallite size[399]. For example, a broad range of 105K was obtained for $Mn_3Zn_{0.6}Ge_{0.4}N$ with an average crystallite size of 54nm, sintered at 700C, and the average linear thermal expansion coefficient was -13.5 x 10^{-6}/K within the negative-expansion operating range.

$Mn_3(Sn,Zn)N$

The thermal expansion of Mn_3ZnN which had been doped with Sn and sintered at 1223K was measured. The compound has a cubic antiperovskite Mn_3CuN-type structure and exhibits a thermal expansion of -19.05 x 10^{-6}/K in the case of $Mn_3Zn_{0.5}Sn_{0.5}N$, with a broad operating width of 60K[400]. When samples of $Mn_3Zn_{1-x}Sn_xN$ (x = 0.05, 0.1, 0.15, 0.2, 0.3) were prepared by solid-state reaction, the transition temperature increased while the difference in volume did not change with increasing tin doping. With x = 0.3, negative thermal expansion behavior occurred at room temperature, indicating that the range of low thermal expansion had broadened to room temperature[401]. The coefficient of thermal expansion in this range increased from 2.35 x 10^{-6} to 8.66 x 10^{-6}/K. With increasing tin content in $Mn_3Zn_{1-x}Sn_xN$ (x = 0.1, 0.2, 0.3, 0.5, 0.8, 1.0) compounds, the thermal expansion behavior changed from positive to negative and then returned to positive near to the magnetic transition temperature range. The magnetic transition temperature meanwhile increased from 185 to 495K[402]. It was noted that the abnormal thermal expansion behavior was related to the number of valence electrons on the zinc site, and that the equivalent effect of three valence electrons was beneficial to the occurrence of negative thermal expansion.

Mn_3ZnN

An isotropic high negative-expansion and high-stiffness material would be able to compensate for the high *positive* thermal expansion of aluminium or a plastic. A thermal expansion coefficient of over -30 x 10^{-6}/K at room temperature can be achieved in Mn_3ZnN-based antiperovskite manganese nitrides by simultaneously substituting carbon and boron for nitrogen, as well as tin for zinc[403]. The resultant nitrides have a larger negative expansion, although the width of the operating-range is comparable to that of other Mn_3CuN-based materials. Study of the magnetic structure of Mn_3ZnN predicts a new magnetic ground state. Comparison of the calculated volumes of various magnetic structures shows that Mn_3ZnN undergoes a volume expansion from the high-temperature paramagnetic phase to the low-temperature antiferromagnetic phase, 5g, and a volume contraction from the 5g phase to the MGS phase; as observed experimentally. Analysis of the exchange parameters between ions shows that the spin coupling between manganese ions is responsible for the sudden expansion and contraction of the Mn_3ZnN volume. Existing nitrogen vacancies in the compound markedly lower the energy of 5g. When the concentration of nitrogen vacancies is sufficiently high, 5g could become the ground state for the defective Mn_3ZnN compound[404]. This might explain why the expected sudden change in volume at about 127K did not occur.

Chapter 11

Metal-Organics

The archetypical negative thermal expansion **Metal-O**rganic **F**ramework material is MOF-5: $Zn_4O(1,4$-benzenedicarboxylate$)_3$. Direct experimental measurement of the thermal expansion, using neutron powder diffraction between 4 and 600K, shows that the linear thermal expansion coefficient is -16 x 10^{-6}/K. Almost all of the low-frequency lattice vibrational modes (below ~23meV) involve the motion of the benzene rings and the ZnO_4 clusters as rigid units and of the carboxyl groups as bridges[405]. These rigid-unit modes exhibit various degrees of phonon softening and are directly responsible for the large negative thermal expansion. Powder diffraction has been used to study the thermal expansion behaviour under helium gas-pressures of 1.7bar, between 100 and 500K, and under 5 to 150bar, between 150 and 300K[406]. The framework exhibits negative thermal expansion at all pressures, but the vibrational motions which cause negative expansion are increasingly damped, with increasing pressure, leading to a decreased negative expansion. X-ray diffraction studies show that twisting, rotation and libration cause the negative thermal expansion of the nanoporous MOF-5. Near-linear lattice contraction is observed between 80 and 500K. The vibrational motions which cause the abnormal expansion are associated with a shortening of certain interatomic distances with increasing temperature. Detailed analysis of atomic position and displacement parameters indicates that two factors cause the effect[407]. One is local twisting and vibrational motion of the carboxylate groups, and the other is a concerted transverse vibration of the linear linkers. Such linkers have been probed, by means of inelastic neutron scattering under vacuum or at a gas pressure of 175bar, in order to distinguish between the pressure and temperature responses of the framework motions. The local structure of the metal centers has been studied by X-ray absorption spectroscopy. This provided direct evidence that the main contributor to the negative expansion is a translational transverse motion of the aromatic ring. This can be damped by applying gas pressure to the sample. The linker motion is highly correlated rather than being local in nature. The lowest-energy motion is a librational motion of the aromatic ring, which does not contribute to negative expansion[408]. The libration is followed by transverse motion of the linker and the carboxylate group. These motions altogether result in a unit-cell contraction with increasing temperature. The mechanism of negative thermal expansion has been studied using a rigid unit mode analysis of the phonons responsible for negative expansion[409]. In addition to confirming the role played by a so-called optical trampoline-mode mechanism, a new acoustic mode mechanism has been identified which makes a major contribution to the macroscopic negative thermal expansion of the framework. The thermodynamic properties and expansion behavior have been investigated within the quasi-harmonic approximation by using density functional theory. From the Grüneisen parameter as a function of temperature, it can be shown that low-frequency phonons are closely associated with the negative thermal expansion[410]. The lowest-frequency phonon was confirmed to be the one that is most responsible for thermal contraction.

Molecular dynamics simulations show that metal-organic frameworks, constructed from octahedral $Zn_4O(CO_2)O$ clusters linked by aromatic carbon-ring structures, lead to negative thermal expansion from 0K up to the melting-point. Thus MOF-C22 contracts volumetrically by 1.9% between 0 and 600K, with a linear expansion coefficient of -11.05 x 10^{-6}/K, as compared with -9.1 x 10^{-6}/K for ZrW_2O_8. A new MOF was developed, using 2-butynediodate linkers, which leads to a negative thermal expansion of 2.2% between 0 and 500K. This behavior arises because thermal motions in the rigid Zn-O clusters, and in the organic moieties linking them, lead to tilting of the linkers by successively larger degrees from their alignment along the unit-cell axes, resulting in decreased cell parameters[411]. When polyethylene polymer was combined with MOF-C10, the volume of the composite remained constant to within 0.059% between 300 and 600K.

The metal-organic framework material, Cu_3(1,3,5-benzenetricarboxylate)$_2$ exhibits negative thermal expansion over a wide temperature range[412]. The behaviour here originates from two coinciding mechanisms: the concerted transverse vibration of triangular organic linkers and the local dynamic distortion of di-nuclear metal centers within the framework lattice. Framework interpenetration has been found to reduce negative thermal expansion in many materials. X-ray diffraction methods have shown that interpenetrated Cu_3(4,4',4''-benzene-1,3,5-triyl-tribenzoate)$_2$ (MOF-14) exhibits an anomalously large negative expansion[413]. Unlike other interpenetrated materials, the large positive thermal expansion of weak interactions that hold the interpenetrating networks together results here in a low-energy shrinkage of the overall framework. This is a new negative-expansion mechanism. Thermally-densified hafnium terephthalate exhibits one of the greatest isotropic negative thermal expansions reported for a metal-organic framework material. The incorporation of correlated vacancy defects into the framework affects the extent of thermal densification and the degree of negative expansion observed in the densified material[414]. This demonstrates that the inclusion of defects is another means of tuning the behaviour of a metal-organic framework.

The properties of the metal-organic framework material, $[NH_2CHNH_2][Er(HCOO)_4]$, have been studied using nano-indentation; showing that the elastic modulus, hardness and yield stress on the $(021)/(02\overline{1})$ facets are 29.8/30.2, 1.80/1.83 and 0.93/1.01GPa, respectively[415]. X-ray diffraction data show that this formate exhibits significant negative thermal expansion along the b-axis. This is explained in terms of a hinge-strut structural scheme. The thermal expansion behavior of a similar metal-organic framework material, $Zn(HCOO)_2 \cdot 2(H_2O)$, has been studied using X-ray diffraction. This formate has a c-axis thermal expansion coefficient of $-26 \times 10^{-6}K$. The large negative-expansion response is again attributed to a hinge-strut framework motion[416]. Nano-indentation here yields elastic moduli and hardness values, for the $(00\overline{2})/(100)/(110)$ facets, of 35.5/35.0/27.1 and 2.04/1.83/0.47GPa, respectively. Three magnesium formate metal-organic framework materials, $[NH_4][Mg(HCOO)_3]$, $[CH_3CH_2NH_3][Mg(HCOO)_3]$ and $[NH_3(CH_2)_4NH_3][Mg_2(HCOO)_6]$, prove to be of interest. Thus $[NH_4][Mg(HCOO)_3]$, is a chiral framework with NH_4^+ cations located in the channels. Above 255K, these cations oscillate quickly between two shallow energy-minima. Below 255K, they suffer two stages of freezing of the oscillations caused by the inner profiles of the channels and producing non-compensated antipolarization. These factors lead to a significant negative thermal expansion. In perovskite $[CH_3CH_2NH_3][Mg(HCOO)_3]$, the orthorhombic phase below 374K comprises ordered $CH_3CH_2NH_3^+$ cations in the cubic cavities of the magnesium formate framework. The structure becomes trigonal above 374K, with trigonally disordered cations. Another phase transition occurs above 426K and the cation changes to a two-fold disordered state. These two transitions are accompanied by negative and positive thermal expansions. In niccolite $[NH_3(CH_2)_4NH_3][Mg_2(HCOO)_6]$, a gradually increasing flipping movement of the middle ethylene of $[NH_3(CH_2)_4NH_3]_2^+$ in the elongated framework cavity finally leads to a phase transition at 412K[417]. The trigonally disordered cations and framework then change and provide the opportunity for a large negative thermal expansion. The negative thermal expansion which is observed along the chain direction in $[Pt(NH_3)_4][PtBr_2(NH_3)_4]_4](HSO_4)_4$ and $[Pt(NH_3)_4][PtCl_2(NH_3)_4]_4](HSO_4)_4$ is explained in terms of the valence fluctuation of platinum atoms[418]. These crystals are essentially ionic systems with mean valences of $Pt^{3+\delta}$ and $Pt^{3-\delta}$, and the valence fluctuation can be treated in two ways: slow or rapid[419]. In both cases, minimization of the free-energy provides the contraction of $Pt^{3-\delta}$-Br or $Pt^{3-\delta}$-Cl distances and negative expansion of the $Pt^{3+\delta}$-Br or $Pt^{3+\delta}$-Cl distance with increasing temperature. In general, the thermal expansion due to the hindered rotation of the ammonium ion or the methyl group is negative at liquid helium temperatures if the tunnelling frequencies of those polyatomic groups are greater than 0.3/cm. The thermal expansion due to the hindered rotation of polyatomic groups with high tunnelling frequencies is much greater than that due to lattice vibrations at very low temperatures, and so the observed thermal expansion should also be negative. When the barriers to rotation are lower than 0.5kcal/mol, the tunnelling frequencies of NH_4^+ and CH_3 are sufficiently large to cause negative thermal expansion[420]. Some non-cubic ammonium compounds with relatively high barriers to rotation have been predicted to exhibit negative thermal expansion along one of the crystallographic axes.

Crystals of (phenylazophenyl)palladium hexafluoroacetylacetonate can transition between five crystal structures which are linked by four phase transformations; one of them being a thermosalient transformation[421]. Physical changes are controlled by a uniaxial negative expansion which is countered by a huge positive axial expansion of 260 x 10^{-6}/K. The volumetric expansion coefficient is about 250 x 10^{-6}/K. The habit plane advances at some 10000 times the rate which is observed in the case of non-thermosalient transitions.

The structure of $[Zn\{CO(NH_2)_2\}_6](NO_3)_2$ was determined at 110 and 250K. It is stabilized by 12 individual, intra-molecular and inter-molecular, hydrogen bonds. An analysis of the thermal expansion tensor, based upon unit cells over a range of 180K, reveals uniaxial compression in the b-axis direction during heating. The hydrogen bonds form layers which are perpendicular to this axis, and the layers are connected by coordinate bonds parallel to the axis[422]. The inter-molecular hydrogen bonds expand during heating, while the coordinate bonds surprisingly contract and there are associated changes in the O-Zn-O angles; producing an overall so-called accordion effect.

Chapter 12

Simple Oxides

Ag_2O, Cu_2O

It is useful to treat these two oxides together, as they possess so many analogous features. The structure of the cuprite-type oxide, Ag_2O, was studied between 5 and 300K using high-resolution synchrotron X-ray and neutron powder diffraction and the material was found to exhibit a marked negative thermal expansion below 200K. The cell volume is only weakly temperature-dependent between 200 and 300K[423]. An apparently first-order transition at 30K is thought to contribute to the negative expansion. Negative thermal expansion also occurs in the analogous cuprite-type oxide, Cu_2O, below 240K and, in Ag_2O, up to its decomposition temperature of 470K[424]. The calculated Grüneisen parameters for Cu_2O have lower negative values than those for Ag_2O, thus resulting in a smaller degree of negative expansion in Cu_2O. Both oxides are framework structures which comprise two interpenetrating networks of metal-sharing Ag_4O or Cu_4O tetrahedra. There is a strong transverse motion of the metal atoms perpendicular to the O-M-O bonds[425]. Analysis of the next-nearest neighbor shell in EXAFS data indicates a differing temperature behaviour of the Ag-Ag or Cu-Cu interactions between metal atoms related to the same framework, and with respect to metal atoms located on different interpenetrating frameworks. Of the 12 Ag-Ag or Cu-Cu next-nearest neighbor pairs, the 6 which are connected via a bridging oxygen atom exhibit negative expansion, while the 6 which lack bridging oxygen atoms exhibit positive expansion. There is a positive expansion of the nearest-neighbor Ag-O or Cu-O pair distance, and a perpendicular-to-parallel anisotropy of relative motion that is much stronger than the anisotropy of the absolute Ag or Cu motion[426]. The Ag-O or Cu-O bond is much stiffer in stretching than in bending. The tension effect giving rise to negative expansion can be related to the anisotropy of the mean-square relative displacements[427]. Calculated pressure-dependences of the phonon modes have been used to calculate the thermal expansion behavior of Ag_2O, thus showing that the low-energy rotational modes of Ag_4O tetrahedra are responsible for the large negative thermal expansion[428]. The temperature-dependence of the phonon spectrum reveals a markedly anharmonic nature of phonon modes with an energy of about 2.4meV[429]. *Ab initio* calculations show that the maximum negative

Grüneisen parameter, which is a measure of the relevant anharmonicity, occurs for the transverse phonon modes that involve bending motions of the Ag_4O tetrahedra.

The negative thermal expansion mechanism in Cu_2O has been characterized by linking various Cu_2O structural-flexibility models to phonon data obtained by using *ab initio* lattice dynamics theory. The low-frequency acoustic modes which are responsible for the negative thermal expansion in this material correspond to vibrations of the rigid O-Cu-O rods. A small contribution also arises from higher-frequency optical modes which correspond to the rotation of rigid and near-rigid Cu_4O tetrahedra and of near-rigid O-Cu-O rods[430]. The negative expansion behaviour was also linked to a ferroelastic phase transition at high pressures. A temperature-dependent pair distribution function analysis showed that copper oxide, at about 200K, exhibited geometrical distortions of the tetrahedral units which were probably related to a change in the solid angles of the polyhedra. Silver oxide, on the other hand, exhibited the same distortions even at 10K[431]. Quantitative refinement confirmed the presence of local distortions of less than 10Å at all temperatures.

Another small group of materials which it is useful to mention as a whole comprises the three magnetic nanocrystals: CuO, MnF_2 and NiO. Evidence is found for negative thermal expansion in CuO and MnF_2 below their magnetic ordering temperatures, but not in NiO. Larger particles of CuO and MnF_2 also exhibit marked magnetostriction, resulting in significantly reduced thermal expansion below their magnetic ordering temperatures. This is not observed in NiO. It is suggested that the negative thermal expansion of CuO and MnF_2 is a general property of nanoparticles, in which there is a strong coupling between magnetism and the lattice[432]. The negative expansion in CuO is four times greater than that observed in ZrW_2O_8.

Mo_4O_{11}

The two-dimensional oxide, η Mo_4O_{11}, exhibits an almost isotropic thermal expansion in the normal phase, but anisotropic negative thermal expansion in the charge density wave phase below 105K. Here, an anomalous expansion occurs along the a-axis[433]. It is concluded that the charge density wave-induced negative expansion is due to structural relaxation of the MoO_6 octahedral layers along the stacking direction, aided by the presence of open spaces around the MoO_4 tetrahedra.

Nb_2O_5

When the thermal expansion coefficients of a Nb_2O_5 monocrystal and of a Nb_2O_5-TiO_2 ceramic are measured between -200 and 500C by using dilatometry, both sorts of specimen exhibit negative thermal expansion[434]. Some contribution to this behavior is attributed to the presence of Magneli phases.

ReO_3

Rhenium trioxide is an electrically conducting oxide of cubic symmetry. The temperature evolution of the lattice parameters shows that it can exhibit negative thermal expansion below room temperature. The transition point from negative to positive thermal expansion depends upon the sample preparation. Monocrystalline samples exhibit the highest transition temperature, 294K, and the largest negative value of the coefficient of thermal expansion, -1×10^{-6}/K. The atomic displacement parameters for the oxygen atoms are strongly anisotropic, even at 15K. This indicates that there is a large contribution of static disorder to the displacement parameters[435]. The temperature evolution of the oxygen displacement parameters for various samples shows that the static disorder contribution is greater for samples with decreased negative thermal expansion behavior. Neutron diffraction studies show that the lattice parameter and the unit-cell volume decrease continuously as the temperature is increased from 2 to about 200K. After reaching minima at about 200K they then increase linearly with increasing temperature up to about 305K[436]. The negative thermal expansion is attributed to rigid antiphase rotations of neighboring ReO_6 octahedra. Further studies revealed that negative thermal expansion again occurs at 600 to 680K[437]. This was attributed to anharmonicity and to an anomalous softening of the M3 phonon mode.

SiO_2

Computer modelling and theoretical analysis have long been used to explain the nearly-zero and slightly-negative coefficients of thermal expansion in β-quartz when well above the α-β transition. Quartz is the archetype of a material possessing a framework of stiff units, the SiO_4 tetrahedra, linked via the very flexible hinges represented by shared oxygen atoms[438]. The results supported the theory that the negative contribution to thermal expansion in framework structures is a geometrical effect due to the rotation of the tetrahedral units, which fold together as they vibrate. The α and β modifications of quartz and cristobalite were again simulated using molecular dynamics methods based upon parameterization of a charge-transfer three-body potential. The simulated α-phase exhibited positive thermal expansion. It was almost zero for β-cristobalite up to 1500K, and slightly negative at higher temperatures. A negative thermal expansion of β-quartz was observed immediately above the α-to-β transition[439]. A detailed analysis of atomic trajectories revealed that the cause of the negative thermal expansion in the high-temperature β-forms was a gradual reactivation of the same displacement mode which allowed the transformation between the α and β modifications. The variation of the crystal structure of quartz between 298 and 1235K has been studied using synchrotron powder X-ray diffraction and Rietveld structure refinement. The polymorphic transformation from $P3_221$ (α-quartz) to $P6_222$ (β-quartz) occurs at 847K. The Si-Si distance in α-quartz increases with temperature because of cation repulsion and so the Si-O-Si angle increases, causing thermal expansion of the framework-structure comprising corner-sharing distorted rigid SiO_4 tetrahedra[440]. The Si-Si distances contract with temperature and cause negative thermal expansion in β-quartz because of increasing thermal librations, of the oxygen atom in the Si-O-Si linkage, which occur nearly perpendicularly to the Si-Si contraction.

In order to provide a suitable substrate for the expansion-proofing of a fiber Bragg grating, a polycrystalline β-quartz solid solution, $Li_2O\text{-}Al_2O_3\text{-}nSiO_2$ (n > 2), was tested. It had a thermal expansion coefficient of between -6.5×10^{-6} and -8.5×10^{-6}/K. This was sufficiently large to ensure total compensation of the Bragg wavelength-shift. Any hysteresis could be decreased by controlling the SiO_2 ratio and the grain size[441]. The temperature dependence of a grating mounted onto the substrate by using an epoxy adhesive decreased to -2.3×10^{-3}nm/K, from 10.0×10^{-3}nm/K. The hysteresis in the Bragg wavelength shift was less than 0.03nm.

V_2O_5

Vanadium pentoxide gels, $V_2O_5 \cdot 1.6H_2O$, produce xerogel layers which have a preferred orientation and an apparent turbostratic stacking of the V_2O_5 ribbons in a direction parallel to the substrate. The distance along the c-axis decreases continuously, with increasing temperature, up to 180C and is attributed to an extrinsic negative thermal expansion mechanism; with a coefficient which can be as high as -1500×10^{-6}/K. Full recovery of the interlayer spacing occurs upon cooling the sample to room temperature in air, whereupon water molecules are re-absorbed[442]. The process is reversible, and the heating can be repeated without losing the negative expansion effect.

ZnO

First-principles calculations have been made of the vibrational and thermal properties of wurtzite and zincblende ZnO with regard to negative thermal expansion. Such behaviour is predicted to occur below 95K and below 84K in the wurtzite and zincblende phases, respectively. The calculated phonon frequencies and mode Grüneisen parameters of the low-energy modes for the wurtzite structure are in good agreement with those determined experimentally; as is the thermal expansion coefficient. The maximum contribution to the negative expansion arises from low-frequency transverse acoustic modes[443]. In the case of the wurtzite structure, the contributions arising from the longitudinal acoustic and lowest-energy optical modes cannot be neglected.

Chapter 13

Mixed Oxides

Negative thermal expansion behavior is found in many oxides when the oxygen or a cation has a coordination number of 2, as in the MO_2, AM_2O_7, AM_2O_7, $A_2M_3O_{12}$, AMO_5 and AO_3 families, with A an octahedral cation and M a tetrahedral cation. Negative expansion is found in all of these, apart from AO_3, where very low expansion is observed for TaO_2F. An open network is required in order to permit the free transverse thermal motion of oxygen, which is deemed to be cause of negative thermal expansion. The openness of the structure can lead to its collapse as the temperature decreases, and there is essentially no thermal expansion below that transition. Such collapse is opposed when there exists enough ionicity in the bonds holding the network together[444]. The openness of the structure also encourages hydration; which again impairs negative thermal expansion. Contraction during heating occurs more frequently in monoclinic minerals than in those of any crystalline group. In six out of seven monoclinic structural types, the deformation consists of a change in the angle of monoclinicity toward 40°; towards higher symmetry[445]. The cause of the negative thermal expansion is shear deformation. Powder neutron diffraction studies, at 30 to 600K, of $CuAlO_2$, $CuInO_2$, $CuLaO_2$, $CuScO_2$, $CuScO_2$ and $AgInO_2$ revealed a negative thermal expansion of the O-Cu-O linkage in every material. This behavior was especially marked in $CuLaO_2$ and $CuScO_2$, where it persisted up to 600K. This negative expansion led to a negative thermal expansion of the c-axis which was moderated by a positive thermal expansion of the M-O bonds. In the $CuMO_2$ series, the negative expansion behavior increased as the M-ion size increased. No negative expansion was found for the O-Ag-O linkage in $AgInO_2$. The compound nevertheless exhibited negative expansion along the c-axis at low temperatures[446]. In the case of $CuLaO_2$, negative expansion occurred along both the a- and c-axes at low temperatures.

Al_2TiO_5-$ZrTiO_4$

High-temperature structural ceramics based upon these oxides possess excellent thermal-shock-resistance[447]. Samples which had been sintered at 1600C exhibited negative thermal expansion up to 1000C, and much lower thermal expansion coefficients, 0.3 x

10^{-6} to 1.3 x 10^{-6}/K, than that of polycrystalline Al_2TiO_5 (1.5 x 10^{-6}/K). These low thermal expansions were however apparently due to microcracking, caused by the large thermal expansion anisotropy of the crystal axes of the Al_2TiO_5.

(Ba,Sr)Zn$_2$SiGeO$_7$

Solid solutions having the compositions, $Ba_{0.5}Sr_{0.5}Zn_2SiGeO_7$ and $BaZn_2Si_{2-x}Ge_xO_7$, were prepared using various x-values. The compounds exhibited very different thermal expansions, due to their differing structures. The $Ba_{0.5}Sr_{0.5}Zn_2SiGeO_7$ solid solutions had a high-temperature $BaZn_2Si_2$-xGexO$_7$ structure and exhibited negative thermal expansion[448]. The structure remained the same up to about x = 1. The low-temperature phase became stable above this value.

BiNiO$_3$

High negative thermal expansions, over a temperature range of about 100K, are obtainable in perovskite oxides which undergo charge-transfer transitions[449]. The present material exhibits a 2.6% volume reduction under pressure. This is due to a Bi/Ni charge transfer that is shifted to ambient pressures by substituting lanthanum for bismuth. Adjusting the proportions of co-existing low- and high-temperature phases leads to a smooth volume-shrinkage during heating. The linear expansion coefficient for $Bi_{0.95}La_{0.05}NiO_3$ is -137 x 10^{-6}/K[450]. A value of -82 x 10^{-6}/K is found between 320 and 380K using dilatometry of a ceramic pellet. Negative thermal expansion of this perovskite, and the high-temperature orthorhombic and low-temperature triclinic phases, were investigated using first-principles calculations. The optimized structures were ferromagnetic, with an orthorhombic configuration, and a G-type antiferromagnetic triclinic phase[451]. A shrinkage of about 1.07% of the cell volume is attributed to the release of stress introduced by changes in the Ni-O and Bi-O bonds; corresponding to changes in the oxidation states of Bi and Ni ions.

(Bi,La)NiO$_3$

The triclinic-to-orthorhombic phase transition, where very highly negative thermal expansion occurred, was studied in $Bi_{1-x}La_xNiO_3$. A charge-transfer transition from $(Bi/La)^{3+}_{0.5}Bi^{5+}_{0.5}Ni^{2+}O_3$ to $(Bi/La)^{3+}Ni^{3+}O_3$, with associated simultaneous structural distortion was observed[452]. Magnetic ordering was present only in the insulating triclinic phase, while the metallic orthorhombic phase was non-magnetic down to 10K. The partial substitution of La^{3+} for bismuth destabilizes the charge disproportionation of bismuth into Bi^{3+} and Bi^{5+}, and an orthorhombic $(Bi,La)^{3+}Ni^{3+}O_3$ state appears at high temperatures. Because of the shortening of the Ni-O bond length, the high-temperature phase has a smaller volume. The coexistence of two phases, of changing relative fraction, in $Bi_{0.95}La_{0.05}NiO_3$ leads to negative thermal expansion; with a dilatometric thermal expansion coefficient of -82 x 10^{-6}/K. This phenomenon is generally observed in $Bi_{1-x}Ln_xNiO_3$. It is also expected to occur in $BiNi_{1-x}M_xO_3$, where M is a trivalent ion, because nickel replacement by other trivalent ions stabilizes the $Bi^{3+}(Ni,M)^{3+}O_3$ state[453]. There is a 50K hysteresis between the heating and cooling cycles of $Bi_{1-x}Ln_xNiO_3$, due to the first-order nature of the transition.

(Bi,Ln)NiO$_3$

The compositional dependence of negative thermal expansion in $Bi_{1-x}Ln_xNiO_3$ (Ln = La, Nd, Eu, Dy) compounds was investigated, showing that all x = 0.05 materials exhibited coefficients of over -70 x 10^{-6}/K; due to intermetallic charge transfer between Bi^{5+} and Ni^{2+}, as above. The temperature range of occurrence of the negative expansion moved upwards with decreasing ionic radius of the Ln^{3+} element[454]. It is suggested that a partial ordering of Bi/Ln is the cause of that movement and of a suppression of the broadening of the transition.

Bi(Ni,Fe)O$_3$

In the case of $BiNi_{1-x}Fe_xO_3$ (x = 0.05, 0.075, 0.10, 0.15), all of the samples exhibit high negative thermal expansion, with coefficients of linear thermal expansion of more than -150 x 10^{-6}/K produced, as usual, by the now familiar charge transfer between Bi^{5+} and Ni^{2+} within a controlled range near to room temperature. As compared with the above $Bi_{1-x}Ln_xNiO_3$ compounds, any problematic thermal hysteresis is suppressed because the random distribution of iron on nickel sites changes the first-order transition to a second-order transition[455]. The coefficient of thermal expansion of $BiNi_{0.85}Fe_{0.15}O_3$ attains -187 x 10^{-6}/K. In applications, an 18vol% addition of this compound negates the thermal expansion of epoxy resin.

(Bi,Pb)NiO$_3$

The valence distribution and local structure of $Bi_{1-x}Pb_xNiO_3$ (x $\leq$ 0.25) were investigated, revealing that disproportionation of the bismuth ions into Bi^{3+} and Bi^{5+} occurred in all of the samples[456]. It was a long-range disproportionation, with distinct crystallographic sites in the P1 triclinic structure for x $\leq$ 0.15. The ordering was short-range for x = 0.20 and 0.25. An intermetallic charge-transfer occurred between Bi^{5+} and Ni^{2+}, leading to a large volume shrinkage, in all of the samples during heating at about 500K.

$Bi_2Sr_2CaCu_2O_8$

When powder and monocrystalline samples of the high-temperature superconductor, $Bi_2Sr_2CaCu_2O_{8+x}$, were studied using X-ray diffraction[457], its thermal expansion coefficient was found to be negative between 160 and 260K.

$(Ca,Ba,Sr)Nb_2O_6$

Polarization Raman spectra measurements of $(Ca_{0.28}Ba_{0.75})_x(Sr_{0.60}Ba_{0.40})_{1-x}Nb_2O_6$ (x = 0.25, 0.50, 0.75) were used to determine the negative thermal expansion change produced by the calcium concentration. Various count-intensities of Raman peaks having the scattering geometry, Z(XY)Z-, and corresponding to the symmetry species, B2, were attributed to the differing vacancy occupations of the A2 sites of the tetragonal tungsten-bronze type of structure. This could result in a change in the short-range molecular force between the ions filling A2 sites, and the Nb-O octahedra[458]. Given that the thermal properties of materials relate directly to vibrations of the crystal lattice, the negative thermal expansion changes of this material were considered to arise from the geometry-changes caused by various calcium concentrations.

Ca_2RuO_4

A huge negative thermal expansion has recently been discovered in this reduced layer ruthenate. The total volume change which is related to negative expansion attains 6.7% according to dilatometry data. This is twice that of the previous record-breaking volume-change. A negative coefficient of linear thermal expansion of -115×10^{-6}/K has been measured between 145 and 345K. This dilatometric result has to be regarded with some suspicion however, as it is too large a negative thermal expansion to be attributable to the known crystallographic cell-volume temperature-variation mechanisms[459]. The very anisotropic expansion of the crystal grains might be due here to conventional microstructural changes in sintered samples.

$Ca_3Mn_2O_7$

There exists a complex competition between lattice modes of different symmetry in this material, leading to phase coexistence over a wide temperature range and to so-called symmetry-trapping of a soft mode[460]. The latter trapping then leads to a uniaxial thermal expansion of -3.6×10^{-6}/K between 250 and 350K.

$Ca_2(Ru,M)O_4$

The Ca_2RuO_4 itself is a Mott insulator with a metal-to-insulator transition at 357K, and antiferromagnetic order at 110K. Slight replacement of ruthenium by a 3d transition-metal ion shifts the first transition above by weakening the orthorhombic distortion and inducing either metamagnetism or a magnetization reversal below the second transition. Such replacement also provokes negative thermal expansion in $Ca_2Ru_{1-x}M_xO_4$ when M is chromium, manganese, iron or copper; the lattice volume expands upon cooling, with a total volume expansion ratio of as much as 1%. The onset of this negative thermal expansion closely parallels the above transition temperatures, thus suggesting an unusual relationship to the electronic properties[461]. The negative expansion occurs first near to room temperature and extends up to 300K.

$(Ca,Sr)_3Mn_2O_7$

The layered perovskite, $Ca_{3-x}Sr_xMn_2O_7$, changes from a material which exhibits uniaxial negative expansion, to one which exhibits positive thermal expansion, as a function of x. The change is related to the presence of two closely competing phases of differing symmetry[462]. The negative thermal expansion is greatest when the solid solution composition is adjusted so as to be as close as possible to this phase region, but disappears suddenly upon passing though the transition, thus suggesting that negative expansion can be controlled by manipulating phase competition.

$Ca_3Ti_2O_7$

A quasi two-dimensional phonon mode has been identified, in this layered perovskite, which has an acoustic branch with quadratic dispersion. This mode exhibits atomic displacements, perpendicular to the layered $[CaTiO_3]_2$ blocks which make up the structure, and a negative Grüneisen parameter. Due to these features, the mode can be exploited so as to obtain unusual membrane effects; including a tunable negative thermal expansion and a pressure-independent thermal softening of the bulk modulus. The negative thermal expansion depends upon the existence of strong intralayer Ti-O covalent bonding and weaker interlayer interactions[463]. This is a distinct departure from the usual negative thermal expansion mechanisms in perovskites, such as rigid-unit modes, structure transformation and electronic/magnetic ordering.

$CuScO_2$

A thermal expansion of -4.0×10^{-6}/K has been found for the apparent Cu-O bond length below 300K, and this was the first observation of marked negative thermal expansion, based upon an O-M-O linkage instead of an M-O-M linkage, occurring in a metal oxide. It is the transverse thermal motion of copper, rather than the transverse thermal motion of oxygen, which causes the negative expansion[464]. The thermal displacement parameter for copper is necessarily very pronounced perpendicular to the linear O-Cu-O linkage.

$CuScO_2$, $CuLaO_2$

Extended X-ray absorption fine-structure measurement at the K-edge of copper in these materials was used to detect the local cause of negative thermal expansion along the c-axis. A positive expansion of the Cu-O bond was found over the entire temperature range[465]. This observation contrasted with a negative expansion of the distance between average atomic positions, as determined by diffraction studies. It was concluded that a weak temperature-dependence of the parallel mean-square relative displacement indicated the existence of a rather stiff Cu-O bond. A large perpendicular value was assumed to reflect an intense relative motion of the copper atom, with respect to oxygen atoms, perpendicular to the c-axis.

$(Er,Sr)NiO_3$

Specimens of $Er_{0.7}Sr_{0.3}NiO_{3-\delta}$, synthesized using a solid-state method, crystallized with an orthorhombic structure, with rod-shaped grains, and exhibited a thermal expansion coefficient of -39.0 x 10^{-6}/K between 655 and 780K[466]. The results of X-ray diffraction, Raman spectroscopy and differential scanning calorimetry studies suggested that the negative thermal expansion behaviour was related to its transformation from orthorhombic to rhombohedral.

$GdInO_3$

Rietveld refinement of X-ray diffraction patterns shows that a negative thermal expansion between 50 and 100K arises from the triangular lattice of Gd^{3+} ions. Calculations suggest that the $GdInO_3$ lattice is highly frustrated at low temperatures. Calculated normal mode frequencies of the gadolinium-related Γ-point phonon modes reveal a significant magneto-elastic coupling in the system[467]. Competition between the magnetic interaction energy and thermal stabilization energy in controlling the interatomic distances is thought to be the cause of the negative thermal expansion.

$LaCu_3Fe_4O_{12}$

Negative and zero thermal expansions near to room temperature can be achieved in cubic A-site ordered perovskites of the form, $LaCu_3Fe_{4-x}Mn_xO_{12}$. First-principles calculations reveal the mechanism behind these effects and the associated electrical and magnetic properties of A-site ordered perovskites. The temperature-sensitive expansion of Cu-O bonds can trigger transformation from Cu^{3+} to Cu^{2+} and cause a covalent state transition of B-site iron from 3 to 3.75. The resultant shrinkage of the Fe-O bonds plays a crucial role in the volume contraction of the oxide at high temperatures[468]. Thus a discontinuous volume change occurs in the parent $LaCu_3Fe_4O_{12}$ oxide due to first-order intermetallic charge transfer: $3Cu^{2+} + 4Fe^{3.75+} \rightleftarrows 3Cu^{3+} + 4Fe^{3+}$. This leads to negative thermal expansion with a linear coefficient of -22 x 10^{-6}/K between 300 and 340K, when x = 0.75, and to an almost zero expansion of -1.1 x 10^{-6}/K, when x = 1, between 240 and 360K[469]. Magnetic susceptibility measurements indicate appreciable broadening of the antiferromagnetic transition as x increases, in line with the relaxation of a first-order electronic phase transition of the parent material.

LiNbO₃

The thermal expansion coefficients of single crystals, with [Li]/([Li]+[Nb]) = 48.56%, were measured between 50 and 1000C. The magnitudes of the anisotropic coefficients varied with temperature in very different ways: whereas the expansion along the x- or y-axes slowly increased monotonically with temperature, that along the z-axis decreased with increasing temperature[470]. At about 650C, the expansion became zero and then took on negative values with further increases in temperature.

NpFeAsO

A neptunium analogue of the tetragonal layered compound, LaFeAsO, was prepared. High-resolution X-ray powder diffraction measurements failed to detect any lattice transformations down to 5K. This was very different to the observed tetragonal-to-orthorhombic distortion seen in rare-earth analogues and which was associated with the stabilization of a spin-density wave on the iron sub-lattice. A marked expansion of the NpFeAsO lattice parameters was instead observed with decreasing temperature below the Neel point[471]. This was associated with a relative volume change of about 0.2% and with Invar behavior between 5 and 20K.

PbFe₁₂Nb₁₂O₃

Synchrotron powder X-ray diffraction data, and Rietveld refinement between 300 and 12K, revealed that the structure remains monoclinic, Cm, down to 12K. The lattice parameters exhibit anomalies at the magnetic transition temperature, due to spin-lattice coupling[472]. The lattice volume exhibits a thermal expansion of -4.64 x 10^{-6}/K below the Neel point.

PbTiO₃

Perovskite-type PbTiO₃ is a negative thermal expansion material between room temperature and its Curie point of 763K. A first-principles study of its anisotropic thermal expansion within the framework of density-functional theory yields the temperature-dependence of the unit-cell volume from 20 to 520K by calculating the minimum total free energy at each temperature point. The negative thermal expansion could be calculated without using empirical parameters[473]. The *ab initio* calculations reveal that the unique behaviour depends upon the phonon vibrations. Such molecular dynamics simulations also show that an appreciable amount of cubic phase exists at about 300K below the Curie point. The quantitative results indicate that, as the temperature increases, the population of the cubic configuration increases and that of the tetragonal configuration decreases[474]. Thermal fluctuation of the cubic configurations in the tetragonal matrix then make a large contribution to the negative thermal expansion because the cubic configuration has a smaller volume and higher entropy than does the tetragonal matrix.

First-principles calculations and quasi-harmonic approximation methods have been used to investigate the thermal expansion properties of the tetragonal phase. The volume of the unit cell exhibits positive thermal expansion in the a-axis direction and negative expansion in the c-axis direction between 0 and 900K[475]. Calculation of the Grüneisen parameters shows that the A1(TO) mode makes the major contribution to the negative expansion, and that this vibrational mode is the lead, titanium and oxygen atoms' relative motion in the c-axis direction; leading to shrinkage of the unit cell in that direction.

The negative thermal expansion of $PbTiO_3$ ferroelectrics can be controlled by substitution. One method is to introduce a cation deficiency at both the Pb^{2+} (A) and Ti^{4+} (B) sites. Comparison of 8% Pb^{2+}-deficient, 2% Ti^{4+}-deficient and pristine samples showed that - whereas the Pb^{2+} deficiency markedly weakened the negative expansion - the effect of the B-site deficiency was negligible[476]. The effects were attributed to the negative-expansion mechanism of spontaneous volume ferro-electrostriction. Cadmium substitution has an unique effect, compared with other A-site substitutions, in that it enhances the negative thermal expansion. The results of structural, density of states and minimum electron density calculations of $Pb_{1-x}Sr_xTiO_3$, $Pb_{1-x}Ba_xTiO_3$ and $Pb_{1-x}Cd_xTiO_3$ super-cells on the basis of chemical bond first-principles calculations demonstrated that the hybridization between Cd-O orbitals was more marked than that between Pb-O orbitals[477]. The bonding between barium or strontium, and oxygen, is almost ionic. Cadmium has an unusual effect in that it enhances the average bulk coefficient of thermal expansion of $PbTiO_3$. Barium and strontium instead reduce the coefficient[478]. It is concluded that covalency of the bonding between the A-site and oxygen in $PbTiO_3$-based materials is responsible for the improved negative thermal expansion. Compounds of the form, $Pb_{1-x}Bi_xTiO_3$ (x = 0.0 to 0.1), have been prepared in order to study the effect of bismuth. The bismuth clearly weakened the tetragonality of the $PbTiO_3$ solid solution, but increased the spontaneous polarization[479]. The enhanced spontaneous polarization and decreased tetragonality both led to a small volume shrinkage with increasing temperature, and the average volumetric thermal expansion coefficient changed from -19.9×10^{-6}/K for pure $PbTiO_3$ to -5.6×10^{-6}/K for $Pb_{0.90}Bi_{0.10}TiO_3$. The Curie point of $Pb_{1-x}Bi_xTiO_3$ was also slightly increased, compared to that of $PbTiO_3$. When compounds of the form, $PbTi_{1-x}Fe_xO_3$ ($0 \leq x \leq 0.15$), were prepared by solid-state reaction, the average volumetric thermal expansion coefficient of the tetragonal phase increased from -19.9×10^{-6}/K for $PbTiO_3$ to -11.3×10^{-6}/K for $PbTi_{0.09}Fe_{0.10}O_3$. The negative thermal expansion and Curie point of $PbTiO_3$ were reduced by introducing iron[480] (table 5). Decreases in the Curie point and c/a-ratio were attributed to reduced spontaneous polarization.

Table 5 Thermal Expansion Coefficients of $PbTi_{1-x}Fe_xO_3$ Compositions.

Composition	Phase	Temperature Range (C)	Coefficient (/K)
$PbTi_{0.95}Fe_{0.05}O_3$	tetragonal	25-475	-14.9×10^{-6}
$PbTi_{0.95}Fe_{0.05}O_3$	cubic	475-750	36.8×10^{-6}
$PbTi_{0.90}Fe_{0.10}O_3$	tetragonal	25-465	-11.3×10^{-6}
$PbTi_{0.90}Fe_{0.10}O_3$	cubic	465-750	37.5×10^{-6}

Another effective way to control the coefficient of thermal expansion from a giant negative value to a near-zero thermal expansion is by adjusting the spontaneous volume ferroelectrostriction in $PbTiO_3$-$(Bi,La)FeO_3$ ferroelectrics[481]. The mechanism of negative thermal expansion has been studied via high-temperature neutron powder diffraction studies of $0.7PbTiO_3$-$0.3BiFeO_3$ and $0.7PbTiO_3$-$0.3Bi(Zn_{1/2}Ti_{1/2})O_3$, where negative-expansion is enhanced or weakened, respectively. Upon increasing the temperature up to the Curie point, the spontaneous polarization displacement of Pb/Bi is weakened in $0.7PbTiO_3$-$0.3BiFeO_3$ but maintained in $0.7PbTiO_3$-$0.3Bi(Zn_{1/2}Ti_{1/2})O_3$. There is an apparent correlation between the tetragonality (c/a-ratio) and degree of spontaneous polarization[482]. Experimental evidence suggests that the spontaneous polarization which originates from Pb/Bi-O hybridization is strongly associated with negative thermal expansion.

When the structures of $(1-x)PbTiO_3$-$xBiFeO_3$ ($x = 0.3$ or 0.6) were investigated by means of neutron powder diffraction, a splitting shift between iron and titanium atoms was found along the c-axis in $0.7PbTiO_3$-$0.3BiFeO_3$. This splitting did not appear in $0.4PbTiO_3$-$0.6BiFeO_3$. The tetragonal phase of $PbTiO_3$-$BiFeO_3$ exhibited a large spontaneous polarization. The negative thermal expansion of $PbTiO_3$ was markedly enhanced over a wide temperature range by the $BiFeO_3$ substitution[483]. The average bulk thermal expansion coefficient of $0.4PbTiO_3$-$0.6BiFeO_3$ was -39.2×10^{-6}/K. Additions of neodymium and samarium have been made to ferroelectric $0.5PbTiO_3$-$0.5BiFeO_3$ in order to modify the negative expansion[484]. The latter was considerably weakened in $0.5PbTiO_3$-$0.5Bi_{0.8}Nd_{0.2}FeO_3$ and $0.5PbTiO_3$-$0.5Bi_{0.8}Sm_{0.2}FeO_3$, as compared with $0.5PbTiO_3$-$0.5BiFeO_3$. The spontaneous volume ferroelectrostriction was strongly related to the square of the spontaneous polarization displacement.

A recent innovation has been that of semiconducting materials having a controllable thermal expansion: such as the ferroelectrics, $(1-x)PbTiO_3$-$xBi(Co_{2/3}Nb_{1/3})O_3$. By adding the latter material, the band-gap is reduced from 2.60 to 2.26eV; thus imparting semiconducting properties to $PbTiO_3$-based ferroelectrics[485]. A gradual transition from negative to positive thermal expansion, -12.3×10^{-6} to 7.4×10^{-6}/K, also occurs.

When the structures of $Pb_{1-x}Cd_xTiO_3$ (x = 0.03 or 0.06) solid solutions were investigated using X-ray Rietveld methods at room temperature, it was found that the spontaneous polarization displacements of lead, cadmium and titanium atoms relative to the oxygen polyhedron surprisingly decreased; although the c/a ratio increased with cadmium doping[486]. The average bulk thermal expansion coefficient decreased from -19.9×10^{-6}/K for pure $PbTiO_3$ to -24.0×10^{-6}/K for $Pb_{0.94}Cd_{0.06}TiO_3$ (table 6). It was suggested that the negative thermal expansion of $PbTiO_3$ might be the result of hybridization between lead and oxygen atoms.

Table 6 Thermal Expansion Coefficients of $Pb_{1-x}Cd_xTiO_3$ Compositions.

Composition	Phase	Temperature Range (C)	Coefficient (/K)
$PbTiO_3$	tetragonal	25-490	-19.9×10^{-6}
$PbTiO_3$	cubic	500-700	37.2×10^{-6}
$Pb_{0.97}Cd_{0.03}TiO_3$	tetragonal	25-470	-21.6×10^{-6}
$Pb_{0.97}Cd_{0.03}TiO_3$	cubic	500-700	38.3×10^{-6}
$Pb_{0.94}Cd_{0.06}TiO_3$	tetragonal	25-470	-24.0×10^{-6}
$Pb_{0.94}Cd_{0.06}TiO_3$	cubic	500-700	36.9×10^{-6}

High-temperature dilatometric studies of $(Pb_{1-x}Ca_x)TiO_3$ (x = 0.35, 0.35, 0.40, 0.45) ferroelectric ceramics revealed the occurrence of negative thermal expansion when x ≤ 0.40. The thermal expansion coefficients which were found for x = 0.30 by using dilatometry and powder X-ray diffraction, were -8.541 x 10^{-6} and -11 x 10^{-6}/K, respectively. The large negative expansions found for x = 0.30 persisted from 70 to 570K. Substitution of Ca^{2+} reduced the negative thermal expansion coefficient of pure $PbTiO_3$, but permitted the creation of strong sintered ceramics[487]. The negative thermal expansion disappeared above the ferroelectric Curie point and was found only in the tetragonal compositions. A study of the effect of fluorine upon $Ca_xPb_{1-x}TiO_3$ showed that the sintering temperature decreased with increasing fluorine content and decreasing CaO content[488]. The fluorine promoted sintering by forming a low melting-point phase in $Ca_xPb_{1-x}TiO_3$. The addition of less than 8mol% of fluorine did not change the crystal phase of $Ca_xPb_{1-x}TiO_3$, but decreased its degree of negative expansion.

The thermal expansion of polymethylmethacrylate-based polymer composites can be reduced by blending the polymer with $PbTiO_3$. The ceramic particles are homogeneously dispersed in the polymer matrix if the composite is prepared via polymerization of a suspension of $PbTiO_3$ in methylmethacrylate monomer[489].

It has been demonstrated that $PbTiO_3$ particles are an effective filler for adjusting the thermal expansion of copper-matrix composites prepared by spark plasma sintering[490]. The addition of the particles effectively reduces the coefficient of thermal expansion of the composite.

PrFeAsO

This is one of the Fe-As-O family of superconductors and, as usual for its rare-earth members, a tetragonal-to-orthorhombic phase transition occurs on cooling below 136K and striped iron magnetism with K= (1,0,1) is detected below about 85K. The ordered iron moment along the a-axis attains a maximum at about 40K, below which an anomalous expansion of the c-axis occurs and results in a thermal volume expansion of - 0.015% at 2K. This effect is usually suppressed in superconductors[491]: the overall negative expansion along the c-axis, from 60K downwards, does not occur in superconducting samples such as those with fluorine doping of the oxygen site. The substitution of ruthenium for iron suppresses moreover the structural and magnetic phase transitions which occur in undoped PrFeAsO. Negative thermal expansion is exhibited by $PrFe_{1-x}Ru_xAsO$ for x-values of between 0.05 and 0.75. As hinted before, it seems that the lack of superconducting behaviour by these materials may be related to the occurrence of negative expansion[492].

R(Fe,Cr)O$_3$

In three members of the perovskite family, RFe$_{0.5}$Cr$_{0.5}$O$_3$ (R = Lu, Yb, Tm), which have an orthorhombic Pbnm structure and are antiferromagnetic at room temperature, there is a progressive spin reorientation on the transition-metal sub-lattice. In the thulium case, a long-range magnetic ordering of the Tm^{3+} sub-lattice is found. No spin-reorientation is observed in the lutetium sample, where magnetization reversal occurs at a compensation temperature of 225K. A clear magnetostrictive effect is observed in samples with ytterbium and thulium, and is associated with a negative thermal expansion[493]. These effects have been attributed to a magnetoelastic effect, produced by repulsion between the magnetic moments of neighboring transition-metal ions.

(Sm,Sr)MnO$_3$

Negative thermal expansion in Sm$_{0.85}$Sr$_{0.15}$MnO$_{3-\delta}$ apparently originates in Sr^{2+} partly replacing Sm^{3+}. The substituted material, with an orthorhombic pbnm structure, relative density of 95.5% and uniform grain size, has a coefficient of thermal expansion of -10.08 x 10^{-6}/K between 360 and 873K. Experimental data suggest that the negative thermal expansion is related to the electron transfer of manganese ions, wherein Mn^{4+} is converted into Mn^{3+} in association with Mn^{3+}O$_6$ octahedral distortions and oxygen defects[494]. The sample volume continually decreases with increasing temperature.

SrCu$_3$Fe$_4$O$_{12}$

First-principles calculations have been used to investigate this A-site-ordered perovskite. It is antiferromagnetic, and its magnetic ordering is C-type with iron ions of mixed valence at the B-site and paramagnetic Cu^{3+} at the A-site. A temperature-induced Cu-Fe inter-site charge transfer occurs which is mediated by the corner-sharing oxygen atoms, while the magnetic interaction changes from antiferromagnetic to ferrimagnetic. The material is thermodynamically stable at low and high temperatures, and its degenerate phonon branches suggest that it is easy to transfer energy between these modes[495]. Its large negative thermal expansion between 170 and 270K is attributed to temperature-induced intermetallic charge transfer, which relaxes the Sr-O and Fe-O bonding units in the oxide. There is charge disproportionation of Fe^{IV} to Fe^{III} and Fe^{V} below 200K[496]. The valence states continuously transform to Sr^{2+}Cu$_3$$^{2.8+}Fe_4$$^{3.4+}O_{12}$ upon cooling to 200K, followed by charge disproportionation to Sr^{2+}Cu$_3$$^{2.8+}Fe_{3.2}$$^{3+}Fe_{0.8}$$^{5+}O_{12}$ at 4K[497]. This establishes the charge-transfer mechanism, and the electronic phase transitions can be distinguished from the first-order charge-transfer phase transitions, 3Cu^{2+} + 4Fe$^{3.75+}$ $\rightleftarrows$ 3Cu^{3+} + 4Fe^{3+} in Ln^{3+}Cu$_3$$^{2+}Fe_4$$^{3.75+}O_{12}$. Direct observation of the negative thermal expansion reveals the occurrence of reversibility and of a small thermal hysteresis, indicating that such expansion can be attributed to a second-order phase transition[498].

Zn₂GeO₄

The molar heat capacity was measured between 0.5 and 400K. At 298.15K, the standard molar heat capacity is 131.86J/Kmol. The standard molar entropy at 298.15K is 145.12J/Kmol. Below room temperature, the average thermal expansion coefficient is - 3.4 x 10^{-6}/K while, above room temperature, it is 3.9 x 10^{-6}/K. The existence of low-energy modes is supported by the excess heat capacity in Zn₂GeO₄ as compared to the sums of the constituent binary oxides[499].

Chapter 14

Miscellaneous Materials

X-ray diffraction measurements of a wide range of materials between 300 and 1K suggest that many of them, such as PrPtBi, $Ce_{0.75}La_{0.25}B_6$ and MgB_2, exhibit negative thermal expansion at low temperatures in spite of their very different crystal structures. It was therefore concluded that negative expansion at low temperatures is common[500]. An explanation can be offered for the negative expansion in terms of Fermi liquid theory.

$Al_{0.8}Cs_{0.7}P_{0.1}Si_{2.1}O_6$

Various pollucite aluminosilicate compounds were synthesized by multi-step heat treatment. These were: V-replaced pollucite, $Cs_{0.7}V_{0.1}Al_{0.8}Si_{2.1}O_6$, P-replaced pollucite, $Cs_{0.7}P_{0.1}Al_{0.8}Si_{2.1}O_6$ and Nb-replaced pollucite, $Cs_{0.7}Nb_{0.1}Al_{0.8}Si_{2.1}O_6$,. The $Cs_{0.7}V_{0.1}Al_{0.8}Si_{2.1}O_6$ had thermal expansion coefficients of 0.9 x 10^{-6}/K between 30 and 500C, -0.9 x 10^{-6}/K between 30 and 150C and -0.4 x 10^{-6}/K between 30 and 100C. Solid-state ^{29}Si MAS NMR spectroscopy showed that the number of Si–O–Si bonds of tetrahedra in the $Cs_{0.7}V_{0.1}Al_{0.8}Si_{2.1}O_6$ framework was larger than that in $Cs_{0.7}P_{0.1}Al_{0.8}Si_{2.1}O_6$ or $Cs_{0.7}Nb_{0.1}Al_{0.8}Si_{2.1}O_6$. The chemical shift for peaks of the ^{29}Si MAS NMR spectrum of $Cs_{0.7}V_{0.1}Al_{0.8}Si_{2.1}O_6$ was larger than those for $Cs_{0.7}P_{0.1}Al_{0.8}Si_{2.1}O_6$ and $Cs_{0.7}Nb_{0.1}Al_{0.8}Si_{2.1}O_6$. This suggested that the bond angle of the tetrahedra was increased by vanadium substitution[501]. Solid-state ^{27}Al MAS NMR spectroscopy data implied that $Cs_{0.7}V_{0.1}Al_{0.8}Si_{2.1}O_6$ comprised a small number of six-fold coordinated aluminium species, resulting in an increase in the number of the Si–O–Si bonds of the tetrahedra of $Cs_{0.7}V_{0.1}Al_{0.8}Si_{2.1}O_6$.

BiOCuSe

The unit cell of this layered oxyselenide contracts during heating from 180 to 200K, and the effect can be increased by imposing a magnetic field. Negative thermal expansion appears between 240 and 260K, with a larger effect, in 4%Cu-deficient $BiOCu_{0.96}Se$. The negative thermal expansion is attributed to the structural distortion which occurs in the ferromagnetic state. This involves magnetic electrons and conduction charges that are linked by weak interlayer ionic Bi-Se bonding and strong in-plane covalent Se-Cu bonding[502]. The distortion can be increased by applying a magnetic field.

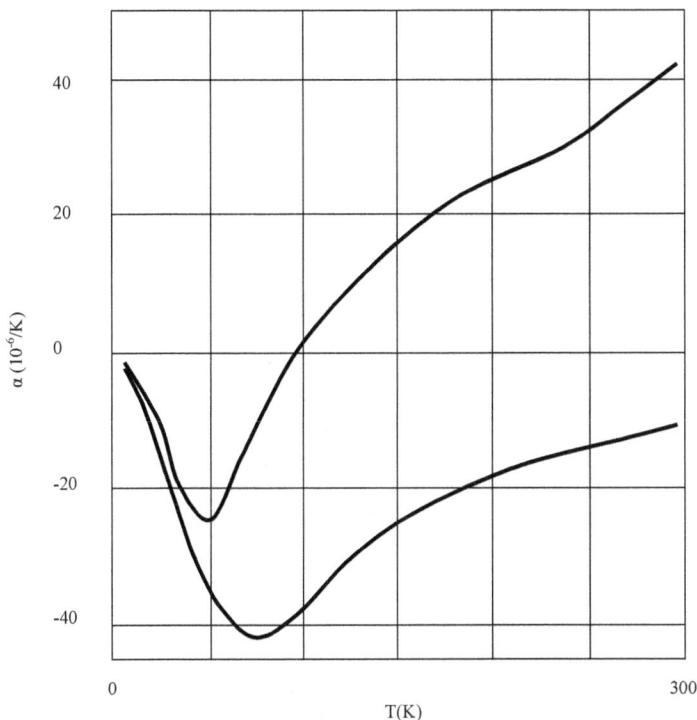

Figure 5 Thermal Expansion of $Ca_{0.80}La_{0.20}Fe_2As_2$
Upper curve: c-axis, lower curve: a-axis

CaCO$_3$

In calcite, the shortest Ca-Ca separation increases with temperature, but the next-nearest Ca-Ca distance, which is equal to the a-axis length, contracts with temperature and causes negative thermal expansion along the a-axis[503]. Thermal librations of the atoms in the rigid CO_3 group increase with temperature along the c-axis.

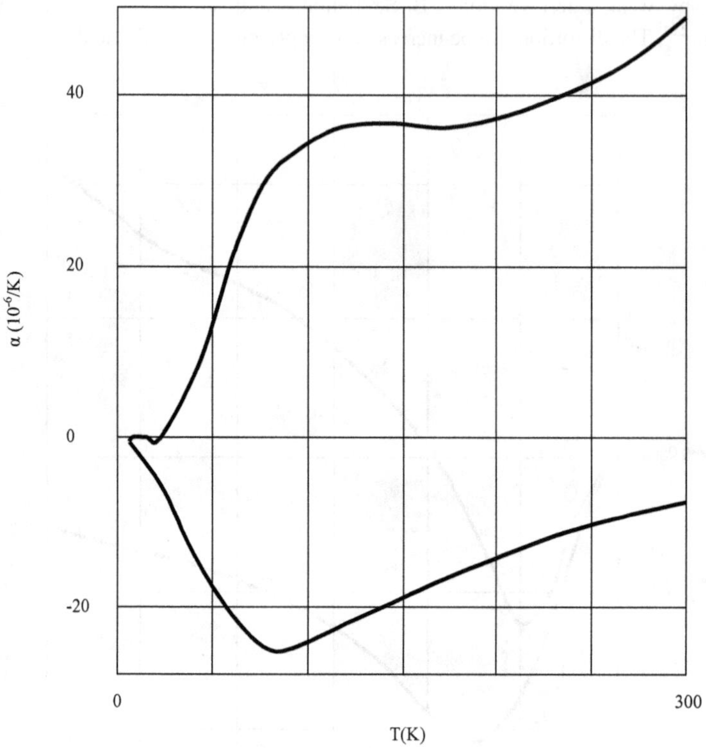

1hr

Figure 6 Thermal Expansion of $Ca_{0.75}La_{0.25}Fe_2As_2$
Upper curve: c-axis, lower curve: a-axis

CaFe₂As₂

Substituting lanthanum for calcium here results in a negative thermal expansion which can be very large; a behavior which is attributed to the close proximity of the material to a structural phase instability[504]. In $Ca_{1-x}La_xFe_2As_2$ (x = 0.15, 0.20, 0.25), the a-axis coefficient of thermal expansion is negative between 0 and 300K for all three samples (figures 5 and 6) although its magnitude is markedly smaller for x = 0.25. The behavior of the c-axis coefficient is quite similar for x = 0.15 and 0.20, but very different for x = 0.25; where no negative expansion is observed along the c-axis[505]. The coefficients can be very large, attaining -41 x 10⁻⁶/K at 73K and -24 x 10⁻⁶/K at 49K for the a- and c-axes, respectively, when x = 0.20. In all of the samples, the value of the coefficient along both axes decreases as the temperature approaches 0K, implying a compliance with the third law of thermodynamics.

CuCl

Extended X-ray absorption fine-structure measurements, and quantitative analysis of the first coordination shell, reveal that the nearest-neighbor Cu-Cl interatomic distance undergoes a large positive expansion. This contrasts with a much weaker negative expansion of the distance between the average atomic positions below 100K. The anisotropy of the relative thermal vibrations, judged by the ratio between the perpendicular and parallel mean-square relative displacements, is quite high, whereas the diffraction thermal factors are isotropic[506]. The relative perpendicular vibrations are related to the tension mechanism, and to the transverse acoustic modes, and are concluded to be responsible for the negative thermal expansion of such zincblende structures.

Cu_2OCl_2

This synthetic melanothallite possesses an orthorhombic, Fddd, structure with a = 7.4691, b = 9.5969, c = 9.700Å, V = 695.3Å^3 and Z = 8. There is a single symmetrically independent copper atom in the structure, coordinated to 2 oxygen and 4 chlorine atoms. The mixed-ligand CuO_2Cl_4 octahedron exhibits a marked Jahn-Teller (4 + 2)-distortion, with 2 oxygen and 2 chlorine atoms located in the equatorial plane, and with 2 chlorine atoms located in apical positions. The overall structure is thus a three-dimensional framework, formed by the cross-linking of chains of edge-sharing CuO_2Cl_2 squares; the chains being parallel to [110] and [110] and linked together by shared oxygen atoms. The structure can also be seen as being an array of OCu_4 oxygen-centered tetrahedra linked, by shared copper corners, into a cuprite-like three-dimensional framework where the cavities of the framework contain chlorine anions. Melanothallite exhibits a thermal expansion along the b-axis of -26.7 x 10^{-6}/K, while the expansion along the a-axis is 50.6 x 10^{-6} + 25.2 x 10^{-9}T/K. This response is attributed to the cross-linking of chains of edge-sharing CuO_2Cl_2 squares: at room temperature, two chains are inclined to each other by about 76°. As the temperature increases, this angle tends toward 90°, and the change in angle is associated with an increase in the a-axis cell parameter and a decrease in the b-axis parameter[507]. This so-called hinge mechanism explains the high anisotropy of the thermal behavior.

$Ge_{31}P_{15}Se_8$

The heat capacity and thermal expansion of this semiclathrate were studied between 2 and 300K, showing that the total heat capacity exceeds 1J/Kmol at low temperatures due to a glass-like contribution arising from irregularities in the structure. A Schottky-like term was attributed to an asymmetry of the guest atom location inside distorted polyhedral cages[508]. A negative thermal expansion below 30K was ascribed to guest-atom transitions between energetically non-equivalent states within the distorted cages of the structure.

$KBe_2BO_3F_2$

The $KBe_2BO_3F_2$ family (with rubidium or cesium interchangeable with potassium) comprises the only known borates which exhibit isotropic area negative thermal expansion over a very wide temperature range. First-principles calculations indicate that the area negative expansion behavior originates mainly from the coordinated distortion of [BeO_3F] tetrahedra in the two-dimensional [$Be_2BO_3F_2$] framework as a function of temperature. The [BO_3] triangles remain essentially rigid[509]. The differing sizes of the effect among the potassium, rubidium and cesium variants are attributed to differences in the interactions between those alkali metal atoms and the [$Be_2BO_3F_2$] layer.

LuB_2

The heat capacities and unit-cell volumes of the lutetium borides, LuB_2 and LuB_4, between 2 and 300K were analysed in terms of the Debye-Einstein approximation. The characteristic temperatures of the Debye and Einstein components of the heat capacity and thermal expansion were thus found, and an anomalous contribution to the thermal characteristics was detected[510]. The latter was attributed to the existence of a two-level sub-system of lutetium ions which arises due to an asymmetry in the manner in which they are surrounded by neighboring boron atoms. The two-level system leads to negative thermal expansion. The latter behaviour has been experimentally observed in LuB_{50}. Analysis of the temperature dependence of the Grüneisen parameter of LuB_{50} suggests that low-frequency oscillations are responsible for the negative expansion[511].

$LiAlSiO_4$

β-eucryptite exhibits anisotropic thermal expansion and one-dimensional superionic conductivity. The anisotropic stress dependence of the phonon spectrum has been calculated in order to deduce the thermal expansion behavior along various axes. The results show that the Grüneisen parameters of the low-energy phonon modes around 10meV have large negative values and control negative thermal expansion behavior at low temperatures along both the a- and c-axes. Anisotropic elasticity, together with anisotropic positive values of the Grüneisen parameters of modes in the range 30 to 70meV range are responsible meanwhile for the thermal expansion at high temperatures. The latter is positive in the ab-plane and negative along the c-axis[512]. Softening of a Γ-point mode at about 2GPa is thought to be related to the high-pressure phase transition. When lithium aluminosilicate substrates of β-eucryptite form are prepared by solid-state sintering, doping with TiO_2 improves their density, strength, hardness and coefficient of thermal expansion[513]. The negative thermal expansion is impaired however if more than 3wt% is added. The coefficient of thermal expansion of sintered lithium aluminosilicate samples has been studied down to cryogenic temperatures[514]. Lithium aluminosilicate $Li_2O:Al_2O_3:SiO_2$ powder precursors, with 1:1:2 and 1:1:3.11 composition ratios were sintered by using the spark plasma technique[515]. The coefficient of thermal expansion was studied down to cryogenic conditions.

$LiBeBO_3$

The rare area negative thermal expansion phenomenon is again observed in this relatively newly evaluated alkali beryllium borate, which comprises $[BeBO_3]_\infty$ double layers that are connected by edge-sharing BeO_4 tetrahedra[516]. The unusual thermal behavior is attributed to the combined vibrational effects of abnormal Be-O structures and of Li^+ cations.

MgB_2

The thermal expansion coefficient of MgB_2 changes from positive to negative upon cooling through the superconducting transition temperature. The Grüneisen function also becomes negative at that temperature, followed by a sharp increase to large positive values at low temperatures[517]. This is attributed to an anomalous coupling between superconducting electrons and low-energy phonons.

Na_3OBr

This soft cubic material has a bulk modulus of 58.6GPa, is stable under pressures of up to 23GPa and exhibits negative thermal expansion between 20 and 55K[518].

Na_4OI_2

This soft tetragonal material has a bulk modulus of 52.0GPa, is stable under pressures of up to 23GPa and - although having a layered structure – exhibits almost isotropic compressibility. Negative thermal expansion is observed between 20 and 80K[519].

$Sn_2P_2S_6$

A thermal expansion of -47 x 10^{-6}/K is exhibited by this non-perovskite lead-free ferroelectric between 243 and 338K[520]. The anomalous negative value is believed to be associated with spontaneous polarization and spontaneous volume ferroelectrostriction.

$Tm_2CrFe_{14.5}Si_{1.5}$

Negative thermal expansion has been measured along the a-axis between 423 and 448K, giving an average linear coefficient of -7.16 x 10^{-6}/K. Between 303 and 398K negative thermal expansion was also found along the c-axis, and the average linear coefficient was -20.4 x 10^{-6}/K. These two results gave an average volume thermal expansion coefficient of -7.94 x 10^{-6}/K between 423 and 448K[521]. The material had an hexagonal Th_2Ni_{17}-type structure between 303 and 623K, and the Curie temperature was 453K.

$ZnCr_2Se_4$

This bond-frustrated material exhibits a marked spin-lattice coupling which is characterized by a large magnetostriction and a negative thermal expansion. A systematic investigation of monocrystalline $ZnCr_2(Se_{1-x}S_x)_4$ (x = 0 to 0.1) shows that, with increasing sulfur content, antiferromagnetic ordering is gradually replaced by a spin-glass state. The temperature range of negative thermal expansion expands, and the magnetostriction gradually subsides[522]. These changes are attributed to an enhancement of antiferromagnetic interaction and bond disorder due to sulfur substitution.

Chapter 15

Semiconductors

According to first-principles calculations, all II-VI semiconductors having a cubic zincblende structure exhibit negative thermal expansion at low temperatures. Negative mode Grüneisen parameters are found for two transverse acoustic branches near to the Brillouin-zone boundaries. Analysis of vibrational modes shows that the librational mode, which causes a bond tension effect due to atomic motions perpendicular to the bonds, can contribute to the negative expansion[523]. With increasing ionic radius and atomic mass, there can be a more covalent bonding, weaker interatomic force constants and lower frequencies of the lattice vibrations.

CdTe

The degree and temperature range of the low-temperature negative thermal expansion of this material are intermediate between those of iso-structural germanium and CuCl. Extended X-ray absorption fine-structure measurements, performed between 19 and 300K, permitted an accurate determination of the bond thermal expansion and of the parallel and perpendicular mean square relative displacements[524]. The positive contributions to thermal expansion, due to bond stretching, and the negative contribution due to tension effects were disentangled and calculated[525]. A comparison with previous extended X-ray absorption fine-structure results for germanium and CuCl showed that correlations exist between various local parameters, and the properties of the negative thermal expansion of tetrahedrally bonded semiconductors.

The luminescence properties of molecular-beam-epitaxially grown layers of (CdMnMg)Te between 2 and 200K show that the anomalous temperature dependence of the lattice constant is related to the excitonic recombination and internal transition of manganese. The temperature behavior of the internal-transition energy is non-monotonic. The existence of a minimum in the photon energy is associated with the change in sign of the thermal expansion coefficient. The decay constants of the internal transition of manganese begin to decrease sharply at the above minimum[526]. By assuming a lattice-constant dependent energy-transfer rate to the infra-red emitting state of the manganese ion, the variation of the lifetimes can be explained by the temperature-dependence of the lattice constant.

GaP

The thermal expansion of the lattice constant of crystal samples was measured between 4.2 and 300K, and a negative thermal expansion coefficient was found at low temperatures[527]. The Grüneisen parameter was negative below 50K.

InSe

Neutron diffraction studies of a layered single crystal and of powders of the γ-polytype, performed between 10 and 300K, show that – between 10 and 50K - the excitation of bending vibrations due to charge density waves alters the phonon spectrum and gives rise to negative thermal expansion in the plane of the layers. This expansion, parallel to the c-axis, was -2.2 x 10^{-6}/K, and characteristic of two-dimensional structures[528]. The average coefficients of thermal expansion between 50 and 300K were 10.48 x 10^{-6}/K, perpendicular to the c-axis, and 12.97 x 10^{-6}/K, parallel to the c-axis..

Si

The phonon dispersion relationship and the Grüneisen constant were calculated by using the adiabatic potential in the bond charge model. The size of the bond charge and its volume change were used as adjustable parameters. The Grüneisen constant for the transverse acoustic mode was positive at small wave-numbers, and negative at large wave-numbers[529]. This behavior arose from the positive volume dependence of the bond charge and was the cause of negative thermal expansion.

SnSe

Inelastic neutron scattering and nuclear resonant inelastic X-ray scattering studies of this Pnma revealed a strong inhomogeneous shift and broadening of the phonon spectrum during heating. First-principles simulations explained its anisotropic thermal expansion and, in particular, a negative thermal expansion within the Sn-Se bilayers[530]. Inclusion of the anisotropic strain dependence of the phonon free energy, together with the electronic ground-state energy, is central to explaining the negative thermal expansion. On the basis of Grüneisen theory, the elastic properties and thermal expansion of the bulk Pnma phase were deduced by means of first-principles calculations. Numerical results indicated that the linear thermal expansion coefficient along the a-axis direction is smaller than that along the b-axis; while that along the c-axis has an appreciable negative value, even at high temperatures. The numerical results were in good agreement with experiment. The

SnSe has a negative Poisson ratio. The two phonon modes which make the most significant contributions to the negative expansion are transverse vibrations perpendicular to the c-axis[531]. Negative thermal expansion can occur even in the presence of all-positive macroscopic Grüneisen parameters.

$TlGaSe_2$

Analysis of the temperature dependences of thermal expansion in crystals having a layered crystalline structure shows that in examples such as graphite, boron nitride, GaSe, GaS and InSe, the negative thermal expansion can be explained in terms of the phonon spectra. In the layer plane of the present material however the negative thermal expansion is the result of negative area compressibility[532]. Its thermal expansion can be controlled by incident light, an external electric field or annealing. With regard to the effect of incident light or of an external electric field, a marked change in the negative linear expansion coefficient has been observed[533]. The two main mechanisms involved are deemed to be the formation of an electret state and a reverse piezoelectric effect. The negative thermal expansion in the layer plane of the semiconductor had previously been attributed to a large Poisson contraction[534].

$ZnCr_2Se_4$

In this bond-frustrated material there is an ideal paramagnetic state above 100K. Below this, ferromagnetic clusters coexist with a paramagnetic state down to the Neel point. By modelling the inverse susceptibility above 100K using a modified paramagnetic Curie-Weiss law, an exponentially changeable exchange integral is deduced[535]. In the face of the variable integral, magnetic exchanges and lattice elastic energy couple with each other via magnetoelastic interaction in the ferromagnetic clusters, such that negative thermal expansion occurs with a loss of exchange energy but a gain in lattice elastic energy.

Chapter 16

Organic Materials

Remembering the example of the anomalous behaviour of the familiar polymer, rubber, one rather expects a negative coefficient of thermal expansion in other polymers to be more likely than in crystalline materials. When thermal expansion coefficients were measured in ultra-oriented polyethylene fibers, prepared by solid-state extrusion at 136C under 2100atm using a drawing ratio of 50, the coefficients of expansion between -150 and +50° perpendicular to the fiber axis were of the order of $100 \times 10^{-6}/K$[536]. In the parallel direction, the values were of the order of $-1 \times 10^{-6}/K$. Negative coefficients were attributed to the existence of highly-extended fully-aligned chains along the fiber. The lateral thermal motion of atoms in an inextensible polymer chain can be expected to cause shortening along the chain direction. This effect could then explain the observed negative thermal expansion along the chain direction of a polymer crystal[537]. Using this assumption, the expansion was calculated for a lattice of parallel linear chains, giving a value of $-13 \times 10^{-6}/K$ for chains having a carbon backbone. This prediction agrees well with data on polyethylene. A molecular mechanics parameter-free model, with 0K potentials, was used to describe this shortening of the unit cell of polyethylene crystals in the chain direction due to an increasing amplitude of torsional vibrations[538]. A self-consistent phonon approach was used to calculate the temperature dependence of the angular displacement correlation function, while taking account of the anharmonicity of the potential. In order to assess the importance of thermal contraction due to chain torsional motion below the lattice melting point, several other models have been proposed. They indicate that reasonably small rms torsional rotation amplitudes can explain the entire negative chain-direction lattice thermal expansion for polyethylene. These conclusions incidentally also apply to elemental selenium and tellurium if time-averaged planar zig-zag and helical models, respectively, are used. In time-averaged planar zig-zag polydiacetylene, a model based upon torsional chain motion induced contraction indicates that significant torsional motion about a C=C bond is necessary in order to explain the entire chain-direction macroscopic thermal expansion. The occurrence of interstitial or vacancy formation contributes to both the lattice and macroscopic expansion coefficients, but can also result in large differences between the latter parameters[539]. In polyethylene single crystals close to the melting point, Renekerdefect formation could result in a negative chain-direction macroscopic expansion coefficient which is several times larger in absolute magnitude than that of the corresponding negative lattice expansion coefficient. Measurements of the thermal expansion parallel to, and perpendicular to, the drawing direction were carried out on a series of oriented polymers with very different crystallinities (0.36 to 0. 81) and drawing ratios (1 to 20) over wide temperature ranges. The perpendicular expansion increased with temperature, while the parallel expansion decreased to values which were typical of those of polymer crystals, -10 x 10^{-6}/K. This was attributed to the constraining effect of the crystalline bridges between the crystalline blocks[540]. In polymers having lower crystallinity, the parallel expansion could be an order-of-magnitude more negative. This was attributed to the rubber-elastic contraction of taut tie-molecules. Because taut tie-molecules and bridges have markedly different effects upon parallel expansion at high temperatures, this permits a rough determination to be made of their relative fractions. The so-called tie-molecules are the polymer chains, found in semicrystalline polymers, which link two or more crystalline lamellae and permit the transfer of force between regions. The behaviour of polyvinylidene fluoride and polyethylene tie molecules was investigated using molecular dynamics simulations. The simulations permitted the measurement of changes in the tie-molecule length during heating. Two molecular-level mechanisms have been identified. One was an entropically driven mechanism which occurs under low applied forces and applies to all flexible polymers[541]. A second mechanism, which is due to conformational changes, is observed in polyvinylidene fluoride but not in polyethylene under intermediate applied forces.

High-performance fiber-reinforced composites which exhibit negative or low thermal expansion have been based upon polyethylene, poly-p-phenylenebenzobisoxazole, aramid or carbon fibers and named Dyneema, Zylon, Technora and Torayca, respectively[542]. The Dyneema and Zylon composites exhibited a significant negative thermal expansion. A porous composite exhibiting planar negative thermal expansion was prepared using multi-material photopolymer additive manufacturing. The internal geometry had been designed by topology optimization, which minimized thermal stresses and maximized stiffness[543]. The resultant structure was then turned into a three-dimensional stereolithography model and assembled. The thermal expansion of the specimen, as measured using laser scanning dilatometry could be greater than -100 x 10^{-6}/K. Hysteresis and fatigue which are associated with large-scale polymer-chain rearrangement can be a problem. Polyarylamide film which contains s-dibenzocyclooctadiene, that can generate completely reversible thermal contraction under low-energy stimulation, exhibits a thermal expansion coefficient of about -1200 x 10^{-6}/K at near to room temperature[544]. Mechanical testing and density-functional theory calculations suggest that there is a conformational change, in the s-dibenzocyclooctadiene component, from a global thermodynamic energy minimum twist-boat configuration to a local minimum chair configuration, and that this is the cause of the abnormal expansion. In-plane and through-thickness measurements of the thermal expansion coefficients of 50% volume fraction glass/polypropylene fibre composites, between room temperature and 120C, show that reproducible results are obtained only between 20 and 75C. In-plane negative values can be obtained in specific directions, leading to large Poisson ratios. The expansion of the matrix depends strongly upon the temperature[545]. Polypropylene does not however provide a matrix which is sufficiently stable to permit the system to exhibit consistently negative values of expansion. Isolated experiments show that a thermal expansion of as much as $-50 \times 10^{-6}/K$ is possible.

Non-polymer organics can also exhibit negative thermal expansion, but a wider range of mechanisms tends to be involved. A supramolecular crystal of 1,4-diazabicyclo[2.2.2]octane hydrochloride exhibits extremely large negative thermal expansions. Its thermal expansion of $-386 \times 10^{-6}/K$ arises due to the structure of honeycomb sheets, made up of hydrogen-bonded H_2O molecules and Cl^- anions. The negative expansion follows H_2O molecule rotation, leading to proton disorder and to a shortening of the hydrogen bonds[546]. An analogous effect occurs in ice. The crystal structures of the simple monoclinic halogen-bonded complexes, pyridine-ICl and pyridine-IBr, yield quadric surfaces - for the thermal expansion tensor - which consist of a hyperboloid of one sheet. Negative thermal expansion occurs parallel to the b-axis, whilst there is a huge expansion approximately parallel to the a-axis. These effects are attributed to a decrease in the strength of the inherently weak $C-H \cdots X$ (X = Br, Cl) hydrogen bond with increasing temperature. This decrease in strength in turn results from a reduction in strength of the halogen bond between the nitrogen atom and the IX component with increasing temperature[547]. Because the I-X bonds are shorter at 298 than at 110K, there is also a smaller partial negative charge on the halogen at 298K; leading to a weaker $C-H \cdots X$ hydrogen bond. The thermal expansion of tienoxolol is anisotropic, and negative along one axis of the unit cell. This expansion behaviour is attributed to its trellis-like supramolecular structure[548]. The negative thermal expansion along one axis of the crystal structures of 1,3,5-benzenetricarboxylic acid and 2,6-dimethylaniline is attributed to the sliding of layers[549]. The negative thermal expansion along the a-axis of pentacene crystals is exceptionally large for a crystalline solid. Between 120 and 413K, the predominant thermal motion is libration of the rigid molecules about their long axes, thus modifying the intermolecular angle which characterises the herring-bone packing within the layers. The herringbone angle increases with temperature to the extent of between 0.3 and 0.6° per 100K and leads to an anisotropic rearrangement of the molecules within the layers: that is, to an expansion in the b-direction and to a marked contraction along the a-direction[550]. A larger herringbone angle improves the co-facial overlap between adjacent parallel molecules and increases the attractive van der Waals forces.

References

[1]   Grima, J.N., Oliveri, L., Ellul, B., Gatt, R., Attard, D., Cicala, G., Recca, G., Physica Status Solidi R, 4[5-6] 2010, 133-135. https://doi.org/10.1002/pssr.201004076

[2]   Grima, J.N., Ellul, B., Attard, D., Gatt, R., Attard, M., Composites Science and Technology, 70[16] 2010, 2248-2252. https://doi.org/10.1016/j.compscitech.2010.05.003

[3]   Lim, T.C., Physica Status Solidi B, 248[1] 2011, 140-147. https://doi.org/10.1002/pssb.200983970

[4]   Wang, Q., Jackson, J.A., Ge, Q., Hopkins, J.B., Spadaccini, C.M., Fang, N.X., Physical Review Letters, 117[17] 2016, 175901. https://doi.org/10.1103/PhysRevLett.117.175901

[5]   Grima, J.N., Ellul, B., Gatt, R., Attard, D., Physica Status Solidi B, 250[10] 2013, 2051-2056. https://doi.org/10.1002/pssb.201384245

[6]   Wu, L., Li, B., Zhou, J., ACS Applied Materials and Interfaces, 8[27] 2016, 17721-17727. https://doi.org/10.1021/acsami.6b05717

[7]   Wang, Y., Lü, E., Wang, Y., Wang, F., Lin, X., Journal of University of Shanghai for Science and Technology, 38[6] 2016, 551-556.

[8]   Qu, J., Kadic, M., Naber, A., Wegener, M., Scientific Reports, 7, 2017, 40643. https://doi.org/10.1038/srep40643

[9]   Grima, J.N., Gatt, R., Ellul, B., Journal of the Chinese Ceramic Society, 37[5] 2009, 743-748.

[10]   Lim, T.C., Journal of Materials Science, 40[12] 2005, 3275-3277. https://doi.org/10.1007/s10853-005-2700-6

[11]   Lakes, R., Applied Physics Letters, 90[22] 2007, 221905. https://doi.org/10.1063/1.2743951

[12]   Dudek, K.K., Attard, D., Caruana-Gauci, R., Wojciechowski, K.W., Grima, J.N., Smart Materials and Structures, 25[2] 2016, 025009. https://doi.org/10.1088/0964-1726/25/2/025009

[13]   Ng, C.K., Saxena, K.K., Das, R., Saavedra Flores, E.I., Journal of Materials Science, 52[2] 2017, 899-912. https://doi.org/10.1007/s10853-016-0385-7

[14]   McNeill, K.G., American Journal of Physics, 28[4] 1960, 375-377. https://doi.org/10.1119/1.1935807

[15] Stepanov, I.A., Materials Letters, 161, 2015, 365-367. https://doi.org/10.1016/j.matlet.2015.08.150

[16] Stepanov, I.A., Journal of Non-Crystalline Solids, 356[23-24] 2010, 1168-1172. https://doi.org/10.1016/j.jnoncrysol.2010.03.013

[17] Stepanov, I.A., Materials Research Innovations, 12[1] 2008, 28-29. https://doi.org/10.1179/143307508X270794

[18] Munn, R.W., Journal of Physics C, 5, 1972, 535-542

[19] Argaman, U., Eidelstein, E., Levy, O., Makov, G., Physical Review B, 94[17] 2016, 174305. https://doi.org/10.1103/PhysRevB.94.174305

[20] Fornasini, P., El All, N.A., Ahmed, S.I., Sanson, A., Vaccari, M., Journal of Physics - Conference Series, 190, 2009, 012025. https://doi.org/10.1088/1742-6596/190/1/012025

[21] Yang, D.S., Lee, J.M., Journal of Physics - Conference Series, 430[1] 2013, 012005. https://doi.org/10.1088/1742-6596/430/1/012005

[22] Fang, H., Dove, M.T., Phillips, A.E., Physical Review B, 89[21] 2014, 214103. https://doi.org/10.1103/PhysRevB.89.214103

[23] Fang, H., Dove, M.T., Physical Review B, 87[21] 2013, 214109. https://doi.org/10.1103/PhysRevB.87.214109

[24] Fang, H., Dove, M.T., Journal of Physics Condensed Matter, 26[11] 2014, 115402. https://doi.org/10.1088/0953-8984/26/11/115402

[25] Sheard, F.W., Physics Letters A, 30[3] 1969, 156-157. https://doi.org/10.1016/0375-9601(69)90907-4

[26] Fetisov, E.P., Khomskii, D.I., Solid State Communications, 56[4] 1985, 403-406. https://doi.org/10.1016/0038-1098(85)90412-0

[27] Stillinger, F.H., Stillinger, D.K., Physica A: Statistical Mechanics and its Applications, 244[1-4] 1997, 358-369. https://doi.org/10.1016/S0378-4371(97)00246-X

[28] Tao, J.Z., Sleight, A.W., Journal of Solid State Chemistry, 173[2] 2003, 442-448. https://doi.org/10.1016/S0022-4596(03)00140-3

[29] Lim, T.C., Journal of Materials Science, 47[1] 2012, 368-373. https://doi.org/10.1007/s10853-011-5806-z

[30] Lim, T.C., Physica Status Solidi B, 250[10] 2013, 2062-2069.

[31]   Hirota, M., Kanno, Y., Optimization and Engineering, 16[4] 2015, 767-809.
https://doi.org/10.1007/s11081-015-9276-z

[32]   Grima, J.N., Farrugia, P.S., Gatt, R., Zammit, V., Proceedings of the Royal Society
A, 463[2082] 2007, 1585-1596. https://doi.org/10.1098/rspa.2007.1841

[33]   Grima, J.N., Bajada, M., Scerri, S., Attard, D., Dudek, K.K., Gatt, R., Proceedings
of the Royal Society A, 471[2179] 2015, 20150188.
https://doi.org/10.1098/rspa.2015.0188

[34]   Schick, J.T., Rappe, A.M., Physical Review B, 93[21] 2016, 214304.
https://doi.org/10.1103/PhysRevB.93.214304

[35]   Rechtsman, M.C., Stillinger, F.H., Torquato, S., Journal of Physical Chemistry A,
111[49] 2007, 12816-12821. https://doi.org/10.1021/jp0768591

[36]   Liu, Z.K., Wang, Y., Shang, S.L., Scripta Materialia, 65[8] 2011, 664-667.
https://doi.org/10.1016/j.scriptamat.2011.07.001

[37]   Kuzkin, V.A., Krivtsov, A.M., Physica Status Solidi B, 252[7] 2015, 1664-1670.
https://doi.org/10.1002/pssb.201451618

[38]   Occhialini, C.A., Handunkanda, S.U., Curry, E.B., Hancock, J.N., Physical
Review B, 95[9] 2017, 094106. https://doi.org/10.1103/PhysRevB.95.094106

[39]   Abdullah, M., International Journal of Modern Physics B, 30[32] 2016, 16502386.
https://doi.org/10.1142/S0217979216502386

[40]   Guillaume, C.E., Comtes Rendus de l'Academie des Sciences, 125, 1897, 235

[41]   Kussmann, A., Rittberg, G.G., Zeitschrift fur Metallkunde, 41, 1950, 470

[42]   Sumiyama, K., Shiga, M., Morioka, M., Nakamura, Y., Journal of Physics F, 9,
1979, 1665 https://doi.org/10.1088/0305-4608/9/8/017

[43]   Lang, M., Schefzyk, R., Steglich, F., Grewe, N., Journal of Magnetism and
Magnetic Materials, 63-64[C] 1987, 79-81. https://doi.org/10.1016/0304-
8853(87)90528-2

[44]   Bastide, C., Lacroix, C., Solid State Communications, 59[3] 1986, 121-125.
https://doi.org/10.1016/0038-1098(86)90192-4

[45]   Lin, Q., Corbett, J.D., Journal of the American Chemical Society, 134[10] 2012,
4877-4884. https://doi.org/10.1021/ja211625w

[46]   Díaz, J., Cid, R., Hierro, A., M Álvarez-Prado, L., Quirós, C., Alameda, J.M.,
Journal of Physics - Condensed Matter, 25[42] 2013, 426002.

https://doi.org/10.1088/0953-8984/25/42/426002

[47]   Christensen, N.E., Boers, D.J., Van Velsen, J.L., Novikov, D.L., Physical Review B, 61[6] 2000, R3764-R3767. https://doi.org/10.1103/PhysRevB.61.R3764

[48]   de Visser, A., Bakker, K., Pierre, J., Physica B, 186-188[C] 1993, 577-579. https://doi.org/10.1016/0921-4526(93)90640-R

[49]   Hao, Y., Gao, Y., Wang, B., Qu, J., Li, Y., Hu, J., Deng, J., Applied Physics Letters, 78[21] 2001, 3277-3279. https://doi.org/10.1063/1.1371968

[50]   Hao, Y., Wang, B., Gao, Y., Chinese Journal of Materials Research, 16[4] 2002, 434-438.

[51]   Hao, Y., Liang, F., He, X., Wu, Y.Z., Qin, Y., Wang, F., Advanced Materials Research, 299-300, 2011, 47-50. https://doi.org/10.4028/www.scientific.net/AMR.299-300.47

[52]   Li, B., Luo, X.H., Wang, H., Ren, W.J., Yano, S., Wang, C.W., Gardner, J.S., Liss, K.D., Miao, P., Lee, S.H., Kamiyama, T., Wu, R.Q., Kawakita, Y., Zhang, Z.D., Physical Review B, 93[22] 2016, 224405. https://doi.org/10.1103/PhysRevB.93.224405

[53]   Liu, Y., Li, W., Huang, R., Xiao, F., Li, L., Rare Metal Materials and Engineering, 44[1] 2015, 228-231.

[54]   Huang, R., Liu, Y., Fan, W., Tan, J., Xiao, F., Qian, L., Li, L., Journal of the American Chemical Society, 135[31] 2013, 11469-11472. https://doi.org/10.1021/ja405161z

[55]   Li, S., Huang, R., Zhao, Y., Li, W., Wang, W., Huang, C., Gong, P., Lin, Z., Li, L., Inorganic Chemistry, 54[16] 2015, 7868-7872. https://doi.org/10.1021/acs.inorgchem.5b00908

[56]   Li, S., Huang, R., Li, W., Wang, W., Zhao, Y., Li, L., Journal of Alloys and Compounds, 646, 2015, 119-123. https://doi.org/10.1016/j.jallcom.2015.05.274

[57]   Li, W., Huang, R., Wang, W., Liu, H., Han, Y., Huang, C., Li, L., Journal of Alloys and Compounds, 628, 2015, 308-310. https://doi.org/10.1016/j.jallcom.2014.11.120

[58]   Li, W., Huang, R., Wang, W., Tan, J., Zhao, Y., Li, S., Huang, C., Shen, J., Li, L., Inorganic Chemistry, 53[11] 2014, 5869-5873. https://doi.org/10.1021/ic500801b

[59]   Li, S., Huang, R., Zhao, Y., Wang, W., Li, L., IOP Conference Series - Materials Science and Engineering, 171[1] 2017, 012154.

https://doi.org/10.1088/1757-899X/171/1/012154

[60]   Fu, B., Zhao, M., Wang, Z., Wang, L., Hao, Y.M., Journal of Functional Materials, 37[11] 2006, 1726-1727+1731.

[61]   Qi, J., Halloran, J.W., Journal of Materials Science, 39[13] 2004, 4113-4118. https://doi.org/10.1023/B:JMSC.0000033391.65327.9d

[62]   Hao, Y., Zhao, M., Zhou, Y., Hu, J., Scripta Materialia, 53[3] 2005, 357-360. https://doi.org/10.1016/j.scriptamat.2005.03.051

[63]   Hao, Y., Zhou, Y., Zhao, M., Journal of Applied Physics, 97[11] 2005, 116102. https://doi.org/10.1063/1.1921334

[64]   Feifei, L., Yanming, H., Zhao, W.Y., Advanced Materials Research, 311-313, 2011, 764-767. https://doi.org/10.4028/www.scientific.net/AMR.311-313.764

[65]   Pandey, A., Mazumdar, C., Ranganathan, R., Tripathi, S., Pandey, D., Dattagupta, S., Applied Physics Letters, 92[26] 2008, 261913. https://doi.org/10.1063/1.2953175

[66]   Lin, J., Tong, P., Zhang, K., Tong, H., Guo, X., Yang, C., Wu, Y., Wang, M., Lin, S., Chen, L., Song, W., Sun, Y., Applied Physics Letters, 109[24] 2016, 241903. https://doi.org/10.1063/1.4972234

[67]   Zhao, Y.Y., Hu, F.X., Bao, L.F., Wang, J., Wu, H., Huang, Q.Z., Wu, R.R., Liu, Y., Shen, F.R., Kuang, H., Zhang, M., Zuo, W.L., Zheng, X.Q., Sun, J.R., Shen, B.G., Journal of the American Chemical Society, 137[5] 2015, 1746-1749. https://doi.org/10.1021/ja510693a

[68]   Shiga, M., Miyake, M., Nakamura, Y., Journal of the Physical Society of Japan, 55[7] 1986, 2290-2295. https://doi.org/10.1143/JPSJ.55.2290

[69]   Xu, K., Li, Z., Liu, E., Zhou, H., Zhang, Y., Jing, C., Scientific Reports, 7, 2017, 41675. https://doi.org/10.1038/srep41675

[70]   Li, Z., Yang, H., Xu, K., Zhang, Y., Zheng, D., Jing, C., Materials Chemistry and Physics, 180, 2016, 156-160. https://doi.org/10.1016/j.matchemphys.2016.05.056

[71]   Li, X., Zheng, Z., Li, J., Li, S., Wei, X., Rare Metal Materials and Engineering, 36[5] 2007, 879-883.

[72]   Jiang, J., Cui, L.S., Zheng, Y.J., Jiang, D.Q., Liu, Z.Y., Zhao, K., Materials Science and Engineering A, 549, 2012, 114-117. https://doi.org/10.1016/j.msea.2012.04.013

[73]   Ahadi, A., Matsushita, Y., Sawaguchi, T., Sun, Q.P., Tsuchiya, K., Acta

Materialia, 124, 2017, 79-92. https://doi.org/10.1016/j.actamat.2016.10.054

[74]   Shi, X., Lian, H., Qi, R., Cui, L., Yao, N., Materials Science and Engineering B, 203, 2016, 1-6. https://doi.org/10.1016/j.mseb.2015.10.005

[75]   Liu, X., Wang, J., Fan, C., Shang, R., Cheng, F., Yuan, B., Song, W., Chen, Y., Liang, E., Chao, M., International Journal of Applied Ceramic Technology, 12[S2] 2015, E28-E33. https://doi.org/10.1111/ijac.12201

[76]   Sanchez, S.I., Menard, L.D., Bram, A., Kang, J.H., Small, M.W., Nuzzo, R.G., Frenkel, A.I., Journal of the American Chemical Society, 131[20] 2009, 7040-7054. https://doi.org/10.1021/ja809182v

[77]   Gschneidner, K.A., Elliott, R.O., Waber, J.T., Acta Metallurgica, 11[8] 1963, 947-955. https://doi.org/10.1016/0001-6160(63)90064-6

[78]   Kuznetsov, A.Y., Dmitriev, V.P., Bandilet, O.I., Weber, H.P., Physical Review B, 68[6] 2003, 064109 https://doi.org/10.1103/PhysRevB.68.064109

[79]   Hao, Y., Zhang, X., Wang, B., Yuang, Y., Wang, F., Journal of Applied Physics, 108[2] 2010, 023915. https://doi.org/10.1063/1.3456444

[80]   Young, L., Gadient, J., Gao, X., Lind, C., Journal of Solid State Chemistry, 237, 2016, 121-128. https://doi.org/10.1016/j.jssc.2016.02.004

[81]   Sumithra, S., Umarji, A.M., Solid State Sciences, 8[12] 2006, 1453-1458. https://doi.org/10.1016/j.solidstatesciences.2006.03.010

[82]   Young, L., Alvarez, P.T., Liu, H., Lind, C., European Journal of Inorganic Chemistry, 2016[8] 2016, 1251-1256.

[83]   Tyagi, A.K., Achary, S.N., Mathews, M.D., Journal of Alloys and Compounds, 339[1-2], 2002, 207-210. https://doi.org/10.1016/S0925-8388(01)02003-5

[84]   Xu, G.F., Liu, Q.Q., Yang, J., Sun, X.J., Cheng, X.N., Ceramics International, 35[8] 2009, 3131-3134. https://doi.org/10.1016/j.ceramint.2009.04.028

[85]   Yang, Y.M., Li, L.C., Feng, M., Chinese Journal of Inorganic Chemistry, 23[3] 2007, 382-386.

[86]   Tyagi, A.K., Achary, S.N., Mathews, M.D., Journal of Alloys and Compounds, 339[1-2] 2002, 207-210. https://doi.org/10.1016/S0925-8388(01)02003-5

[87]   Chai, F.T., Yue, J.L., Qiu, W.J., Guo, H.B., Chen, L.J., Shi, S.Q., Acta Physica Sinica, 65[5] 2016, 056501

[88]   Liu, Q., Yang, J., Cheng, X., Advanced Materials Research, 287-290, 2011, 373-

376. https://doi.org/10.4028/www.scientific.net/AMR.287-290.373

[89] Tyagi, A.K., Achary, S.N., Mathews, M.D., Journal of Alloys and Compounds, 339[1-2] 2002, 207-210. https://doi.org/10.1016/S0925-8388(01)02003-5

[90] Marinkovic, B.A., Ari, M., Jardim, P.M., de Avillez, R.R., Rizzo, F., Ferreira, F.F., Thermochimica Acta, 499[1-2] 2010, 48-53. https://doi.org/10.1016/j.tca.2009.10.021

[91] Xiao, X.L., Wu, M.M., Peng, J., Cheng, Y.Z., Hu, Z.B., Key Engineering Materials, 368-372[2] 2008, 1662-1664. https://doi.org/10.4028/www.scientific.net/KEM.368-372.1662

[92] Evans, J.S.O., Mary, T.A., International Journal of Inorganic Materials, 2[1] 2000, 143-151. https://doi.org/10.1016/S1466-6049(00)00012-X

[93] Li, J., Cheng, X.N., Zhu, J.J., Applied Mechanics and Materials, 320, 2013, 181-184. https://doi.org/10.4028/www.scientific.net/AMM.320.181

[94] Zhu, J., Yang, J., Cheng, X., Journal of the Chinese Ceramic Society, 39[9] 2011, 1456-1461.

[95] Wu, M.M., Peng, J., Han, S.B., Hu, Z.B., Liu, Y.T., Chen, D.F., Ceramics International, 38[8] 2012, 6525-6529. https://doi.org/10.1016/j.ceramint.2012.05.033

[96] Paraguassu, W., Maczka, M., Filho, A.G.S., Freire, P.T.C., Melo, F.E.A., Filho, J.M., Hanuza, J., Vibrational Spectroscopy, 44[1] 2007, 69-77. https://doi.org/10.1016/j.vibspec.2006.08.006

[97] Yang, H.T., Lin, W.L., Shang, F.L., Huang, Y.H., Ling, G., Advanced Materials Research, 79-82, 2009, 1567-1570. https://doi.org/10.4028/www.scientific.net/AMR.79-82.1567

[98] Marinkovic, B.A., Jardim, P.M., De Avillez, R.R., Rizzo, F., Solid State Sciences, 7[11] 2005, 1377-1383. https://doi.org/10.1016/j.solidstatesciences.2005.08.012

[99] Marinkovic, B.A., Ari, M., De Avillez, R.R., Rizzo, F., Ferreira, F.F., Miller, K.J., Johnson, M.B., White, M.A., Chemistry of Materials, 21[13] 2009, 2886-2894. https://doi.org/10.1021/cm900650c

[100] Wang, L., Wang, F., Yuan, P.F., Sun, Q., Liang, E.J., Jia, Y., Guo, Z.X., Materials Research Bulletin, 48[7] 2013, 2724-2729. https://doi.org/10.1016/j.materresbull.2013.04.001

[101] Romao, C.P., Miller, K.J., Johnson, M.B., Zwanziger, J.W., Marinkovic, B.A.,

White, M.A., Physical Review B, 90[2] 2014, 024305.
https://doi.org/10.1103/PhysRevB.90.024305

[102] Wu, M.Y., Wang, L., Jia, Y., Guo, Z.X., Sun, Q., AIP Advances, 5[2] 2015,
027126. https://doi.org/10.1063/1.4913361

[103] Zhou, Y., Zhou, W., Wang, H., Luo, F., Zhu, D., Journal of Polymer Research,
22[7] 2015, 138 https://doi.org/10.1007/s10965-015-0775-6

[104] Zhou, C., Zhang, Q., Liu, S., Luo, B., Yi, E., Tian, E., Li, G., Li, L., Wu, G.,
Physical Chemistry Chemical Physics, 19[19] 2017, 11778-11785.
https://doi.org/10.1039/C7CP00676D

[105] Xiao, X.L., Wu, M.M., Peng, J., Cheng, Y.Z., Hu, Z.B., Key Engineering
Materials, 368-372[2] 2008, 1662-1664.
https://doi.org/10.4028/www.scientific.net/KEM.368-372.1662

[106] Yang, H.T., Lin, W.L., Shang, F.L., Huang, Y.H., Ling, G., Advanced Materials
Research, 79-82, 2009, 1567-1570.
https://doi.org/10.4028/www.scientific.net/AMR.79-82.1567

[107] Song, W.B., Liang, E.J., Liu, X.S., Li, Z.Y., Yuan, B.H., Wang, J.Q., Chinese
Physics Letters, 30[12] 2013, 126502. https://doi.org/10.1088/0256-
307X/30/12/126502

[108] Guzmn-Afonso, C., Gonzlez-Silgo, C., Gonzlez-Platas, J., Torres, M.E., Lozano-
Gorrín, A.D., Sabalisck, N., Snchez-Fajardo, V., Campo, J., Rodríguez-Carvajal,
J., Journal of Physics - Condensed Matter, 23[32] 2011, 325402.
https://doi.org/10.1088/0953-8984/23/32/325402

[109] Ge, X.H., Mao, Y.C., Li, L., Li, L.P., Yuan, N., Cheng, Y.G., Guo, J., Chao, M.J.,
Liang, E.J., Chinese Physics Letters, 33[4] 2016, 046503.
https://doi.org/10.1088/0256-307X/33/4/046503

[110] Liu, Q., Cheng, X., Yang, J., Sun, X., Materials Science and Engineering B,
177[2] 2012, 263-268. https://doi.org/10.1016/j.mseb.2011.12.020

[111] Lind, C., Wilkinson, A.P., Rawn, C.J., Payzant, E.A., Journal of Materials
Chemistry, 11[12] 2001, 3354-3359. https://doi.org/10.1039/b104283c

[112] Liu, Q., Yang, J., Sun, X., Cheng, X., Rare Metal Materials and Engineering,
35[S2] 2006, 83-86.

[113] Stevens, R., Linford, J., Woodfield, B.F., Boerio-Goates, J., Lind, C., Wilkinson,
A.P., Kowach, G., Journal of Chemical Thermodynamics, 35[6] 2003, 919-937.

https://doi.org/10.1016/S0021-9614(03)00050-8

[114] Allen, S., Evans, J.S.O., Physical Review B, 68[13] 2003, 1341011-1341013. https://doi.org/10.1103/PhysRevB.68.134101

[115] Mittal, R., Chaplot, S.L., Schober, H., Kolesnikov, A.I., Loong, C.K., Lind, C., Wilkinson, A.P., Physical Review B, 70[21] 2004, 214303, 1-6.

[116] Lind, C., VanDerveer, D.G., Wilkinson, A.P., Chen, J., Vaughan, M.T., Weidner, D.J., Chemistry of Materials, 13[2] 2001, 487-490. https://doi.org/10.1021/cm000788k

[117] Liu, Q., Fan, C., Liu, J., Sun, X., Cheng, X., Li, H., Journal of Sol-Gel Science and Technology, 76[2] 2015, 279-288. https://doi.org/10.1007/s10971-015-3775-4

[118] Zhang, N., Zhou, W., Chao, M., Mao, Y., Guo, J., Li, Y., Feng, D., Liang, E., Ceramics International, 41, 2015, 15170-15175. https://doi.org/10.1016/j.ceramint.2015.08.090

[119] Li, T., Fu, X., Chang, D., Sun, Q., Wang, F., AIP Advances, 7[3] 2017, 035202. https://doi.org/10.1063/1.4977874

[120] Amos, T.G., Yokochi, A., Sleight, A.W., Journal of Solid State Chemistry, 141[1] 1998, 303-307. https://doi.org/10.1006/jssc.1998.8045

[121] Amos, T.G., Sleight, A.W., Journal of Solid State Chemistry, 160[1] 2001, 230-238. https://doi.org/10.1006/jssc.2001.9227

[122] Woodcock, D.A., Lightfoot, P., Smith, R.I., Materials Research Society Symposium - Proceedings, 547, 1999, 191-196. https://doi.org/10.1557/PROC-547-191

[123] Wallez, G., Clavier, N., Dacheux, N., Bregiroux, D., Materials Research Bulletin, 46[11] 2011, 1777-1780. https://doi.org/10.1016/j.materresbull.2011.08.003

[124] Orlova, A.I., Kemenov, D.V., Petkov, V.I., Zharinova, M.V., Kazantsev, G.N., Samoïlov, S.G., Kurazhkovskaya, V.S., High Temperatures - High Pressures, 34[3] 2002, 315-322. https://doi.org/10.1068/htjr025

[125] Seo, D.K., Whangbo, M.H., Journal of Solid State Chemistry, 129[1] 1997, 160-163. https://doi.org/10.1006/jssc.1997.7301

[126] Cetinkol, M., Wilkinson, A.P., Lee, P.L., Journal of Solid State Chemistry, 182[6] 2009, 1304-1311. https://doi.org/10.1016/j.jssc.2009.02.029

[127] Ge, X., Liu, X., Cheng, Y., Yuan, B., Chen, D., Chao, M., Guo, J., Wang, J., Liang, E., Journal of Applied Physics, 120[20] 2016, 205101.

https://doi.org/10.1063/1.4968546

[128] Shang, R., Hu, Q., Liu, X., Liang, E., Yuan, B., Chao, M., International Journal of Applied Ceramic Technology, 10[5] 2013, 849-856. https://doi.org/10.1111/j.1744-7402.2012.02787.x

[129] Watanabe, H., Tani, J., Kido, H., Mizuuchi, K., Materials Science and Engineering A, 494[1-2] 2008, 291-298. https://doi.org/10.1016/j.msea.2008.04.037

[130] Cetinkol, M., Wilkinson, A.P., Solid State Communications, 149[11-12] 2009, 421-424. https://doi.org/10.1016/j.ssc.2009.01.002

[131] Yuan, B.H., Chen, Y.G., Zhang, Q.L., Chen, L.L., Liu, X.S., Ceramics International, 43[9] 2017, 6831-6835. https://doi.org/10.1016/j.ceramint.2017.02.102

[132] Isobe, T., Umezome, T., Kameshima, Y., Nakajima, A., Okada, K., Materials Research Bulletin, 44[11] 2009, 2045-2049. https://doi.org/10.1016/j.materresbull.2009.07.020

[133] Shi, X.W., Lian, H., Yan, X.S., Qi, R., Yao, N., Li, T., Materials Chemistry and Physics, 179, 2016, 72-79. https://doi.org/10.1016/j.matchemphys.2016.05.011

[134] Zhang, N., Li, L., Wu, M., Li, Y., Feng, D., Liu, C., Mao, Y., Guo, J., Chao, M., Liang, E., Journal of the European Ceramic Society, 36 [11] 2016, 2761-2766. https://doi.org/10.1016/j.jeurceramsoc.2016.04.030

[135] Cheng, Y., Liang, Y., Ge, X., Liu, X., Yuan, B., Guo, J., Chao, M., Liang, E., RSC Advances, 6[59] 2016, 53657-53661. https://doi.org/10.1039/C6RA09666B

[136] Cheng, Y., Liang, Y., Mao, Y., Ge, X., Yuan, B., Guo, J., Chao, M., Liang, E., Materials Research Bulletin, 85, 2017, 176-180. https://doi.org/10.1016/j.materresbull.2016.09.008

[137] Wang, J., Deng, J., Yu, R., Chen, J., Xing, X., Dalton Transactions, 40[13] 2011, 3394-3397. https://doi.org/10.1039/c0dt01562h

[138] Salke, N.P., Gupta, M.K., Rao, R., Mittal, R., Deng, J., Xing, X., Journal of Applied Physics, 117[23] 2015, 235902. https://doi.org/10.1063/1.4922744

[139] Wang, X., Huang, Q., Deng, J., Yu, R., Chen, J., Xing, X., Inorganic Chemistry, 50[6] 2011, 2685-2690. https://doi.org/10.1021/ic200003n

[140] Ge, X., Mao, Y., Liu, X., Cheng, Y., Yuan, B., Chao, M., Liang, E., Scientific Reports, 6, 2016, 24832. https://doi.org/10.1038/srep24832

[141] Yamamura, Y., Horikoshi, A., Yasuzuka, S., Saitoh, H., Saito, K., Dalton

Transactions, 40[10] 2011, 2242-2248. https://doi.org/10.1039/c0dt01087a

[142] Hisashige, T., Yamaguchi, T., Tsuji, T., Yamamura, Y., Journal of the Ceramic Society of Japan, 114[1331] 2006, 607-611. https://doi.org/10.2109/jcersj.114.607

[143] Liu, Q., Yang, J., Rong, X., Sun, X., Cheng, X., Tang, H., Li, H., Materials Characterization, 96, 2014, 63-70. https://doi.org/10.1016/j.matchar.2014.07.001

[144] Liu, Q., Cheng, X., Sun, X., Yang, J., Li, H., Journal of Sol-Gel Science and Technology, 72[3] 2014, 502-510. https://doi.org/10.1007/s10971-014-3465-7

[145] Mittal, R., Chaplot, S.L., Physical Review B, 78[17] 2008, 174303 https://doi.org/10.1103/PhysRevB.78.174303

[146] Sahoo, P.P., Sumithra, S., Madras, G., Guru Row, T.N., Inorganic Chemistry, 50[18] 2011, 8774-8781. https://doi.org/10.1021/ic201224g

[147] Yanase, I., Kojima, T., Kobayashi, H., Solid State Communications, 151[8] 2011, 595-598. https://doi.org/10.1016/j.ssc.2011.02.007

[148] Korthuis, V., Khosrovani, N., Sleight, A.W., Roberts, N., Dupree, R., Warren, W.W., Chemistry of Materials, 7[2] 1995, 412-417. https://doi.org/10.1021/cm00050a028

[149] Zhang, M., Mao, Y., Guo, J., Zhou, W., Chao, M., Zhang, N., Yang, M., Kong, X., Kong, X., Liang, E., RSC Advances, 7[7] 2017, 3934-3940. https://doi.org/10.1039/C6RA26923K

[150] Liu, Q., Yang, J., Sun, X., Cheng, X., Tang, H., Li, H., Applied Surface Science, 313, 2014, 41-47. https://doi.org/10.1016/j.apsusc.2014.05.120

[151] Li, J., Zhu, J.J., Cheng, X.N., Journal of Functional Materials, 44[8] 2013, 1065-1068+1072.

[152] Woodcock, D.A., Lightfoot, P., Ritter, C., Journal of Solid State Chemistry, 149[1] 2000, 92-98. https://doi.org/10.1006/jssc.1999.8502

[153] Cheng, X., Liu, H., Zhang, Z., Qu, Z., Journal of Vacuum Science and Technology, 28[3] 2008, 252-255.

[154] Paraguassu, W., Maczka, M., Filho, A.G.S., Freire, P.T.C., Melo, F.E.A., Filho, J.M., Hanuza, J., Vibrational Spectroscopy, 44[1] 2007, 69-77. https://doi.org/10.1016/j.vibspec.2006.08.006

[155] Yang, Y.M., Li, L.C., Feng, M., Chinese Journal of Inorganic Chemistry, 23[3] 2007, 382-386.

[156] Cao, W., Li, Q., Lin, K., Liu, Z., Deng, J., Chen, J., Xing, X., RSC Advances, 6[98] 2016, 96275-96280. https://doi.org/10.1039/C6RA21136D

[157] Sumithra, S., Tyagi, A.K., Umarji, A.M., Materials Science and Engineering B, 116[1] 2005, 14-18. https://doi.org/10.1016/j.mseb.2004.08.015

[158] Peng, J., Liu, X.Z., Guo, F.I., Han, S.B., Liu, Y.T., Chen, D.F., Zhao, X.H., Hu, Z.B., Transactions of Nonferrous Metals Society of China, 19[6] 2009, 1623-1627. https://doi.org/10.1016/S1003-6326(09)60079-0

[159] Wang, J., Chen, Q., Liang, E.J., 2011 International Workshop on Engineering Application Research, WEAR 2011, 114-118.

[160] Liu, H., Wang, G., Zhang, Z., Pan, K., Zeng, X., Ceramics International, 40[9A] 2014, 13855-13859. https://doi.org/10.1016/j.ceramint.2014.05.103

[161] Liu, H., Wang, G., Zhang, Z., Pan, K., Ma, J., Zeng, X., Journal of the Chinese Ceramic Society, 42[6] 2014, 718-722.

[162] Yamamura, Y., Nakajima, N., Tsuji, T., Koyano, M., Iwasa, Y., Katayama, S., Saito, K., Sorai, M., Physical Review B, 66[1] 2002, 143011-143019. https://doi.org/10.1103/PhysRevB.66.014301

[163] Evans, J.S.O., Mary, T.A., Vogt, T., Subramanian, M.A., Sleight, A.W., Chemistry of Materials, 8[12] 1996, 2809-2823. https://doi.org/10.1021/cm9602959

[164] Gallington, L.C., Chapman, K.W., Morelock, C.R., Chupas, P.J., Wilkinson, A.P., Journal of Applied Physics, 115[5] 2014, 053512. https://doi.org/10.1063/1.4864258

[165] Mittal, R., Chaplot, S.L., Kolesnikov, A.I., Loong, C.K., Mary, T.A., Physical Review B, 68[5] 2003, 054302. https://doi.org/10.1103/PhysRevB.68.054302

[166] Yamamura, Y., Nakajima, N., Tsuji, T., Iwasa, Y., Saito, K., Sorai, M., Solid State Communications, 121[4] 2002, 213-217. https://doi.org/10.1016/S0038-1098(01)00466-5

[167] Jorgensen, J.D., Hu, Z., Short, S., Sleight, A.W., Evans, J.S.O., Journal of Applied Physics, 89[6] 2001, 3184-3188. https://doi.org/10.1063/1.1347412

[168] Li, T., Ge, X., Liu, X., Cheng, Y., Liu, Y., Yuan, H., Li, S., Liu, Y., Guo, J., Sun, Q., Li, Y., Liang, E., Materials Express, 6[6] 2016, 515-520. https://doi.org/10.1166/mex.2016.1337

[169] Nakajima, N., Yamamura, Y., Tsuji, T., Journal of Thermal Analysis and Calorimetry, 70[2] 2002, 337-344. https://doi.org/10.1023/A:1021647717948

[170] Suzuki, T., Omote, A., Journal of the American Ceramic Society, 87[7] 2004, 1365-1367. https://doi.org/10.1111/j.1151-2916.2004.tb07737.x

[171] Gindhart, A.M., Lind, C., Green, M., Journal of Materials Research, 23[1] 2008, 210-213. https://doi.org/10.1557/JMR.2008.0013

[172] Guo, F., Chen, X., Deng, X., Ma, H., Yang, X., Zhao, X., Journal of Solid State Chemistry, 196, 2012, 119-124. https://doi.org/10.1016/j.jssc.2012.05.040

[173] Liu, H., Ma, J., Zhang, Z., Zhu, J., Zeng, X., Chen, X., Journal of the Chinese Ceramic Society, 42[9] 2014, 1116-1120.

[174] Liu, H., Zhang, Z., Ma, J., Jun, Z., Zeng, X., Ceramics International, 41[8] 2015, 10469, 9873-9877.

[175] Karna, S.K., Wang, C.W., Sankar, R., Avdeev, M., Singh, A., Panneer Muthuselvam, I., Singh, V.N., Guo, G.Y., Chou, F.C., Physical Review B, 92[1] 2015, 014413. https://doi.org/10.1103/PhysRevB.92.014413

[176] Sumithra, S., Tyagi, A.K., Umarji, A.M., Materials Science and Engineering B, 116[1] 2005, 14-18. https://doi.org/10.1016/j.mseb.2004.08.015

[177] Forster, P.M., Yokochi, A., Sleight, A.W., Journal of Solid State Chemistry, 140[1] 1998, 157-158. https://doi.org/10.1006/jssc.1998.7967

[178] Peng, J., Liu, X.Z., Guo, F.L., Han, S.B., Liu, Y.T., Chen, D.F., Hu, Z.B., International Journal of Minerals, Metallurgy and Materials, 17[6] 2010, 786-790. https://doi.org/10.1007/s12613-010-0390-9

[179] Evans, J.S.O., Mary, T.A., Sleight, A.W., Journal of Solid State Chemistry, 133[2] 1997, 580-583. https://doi.org/10.1006/jssc.1997.7605

[180] Zhu, J.J., Cheng, X.N., Yang, J., Journal of Functional Materials, 42[3] 2011, 553-556.

[181] Evans, J.S.O., Mary, T.A., Sleight, A.W., Journal of Solid State Chemistry, 137[1] 1998, 148-160. https://doi.org/10.1006/jssc.1998.7744

[182] Liu, H., Yang, L., Zhang, Z., Pan, K., Zhang, F., Cheng, H., Zeng, X., Chen, X., Ceramics International, 42[7] 2016, 8809-8814. https://doi.org/10.1016/j.ceramint.2016.02.124

[183] Zhang, Z.P., Liu, H.F., Pan, K.M., Chen, X.B., Zeng, X.H., Journal of Inorganic Materials, 30[12] 2015, 1278-1282. https://doi.org/10.15541/jim20150219

[184] Weller, M.T., Henry, P.F., Wilson, C.C., Journal of Physical Chemistry B, 104[51] 2000, 12224-12229. https://doi.org/10.1021/jp0030037

[185] Secco, R.A., Liu, H., Imanaka, N., Adachi, G., Journal of Materials Science Letters, 20[14] 2001, 1339-1340. https://doi.org/10.1023/A:1010967021588

[186] Zhou, Y., Neiman, A., Adams, S., Physica Status Solidi B, 248[1] 2011, 130-135. https://doi.org/10.1002/pssb.201083969

[187] Yamamura, Y., Ikeuchi, S., Saito, K., Chemistry of Materials, 21[13] 2009, 3008-3016. https://doi.org/10.1021/cm900965p

[188] Liu, Q., Yang, J., Cheng, X., Liang, G., Sun, X., Ceramics International, 38[1] 2012, 541-545. https://doi.org/10.1016/j.ceramint.2011.07.041

[189] Liu, F.S., Chen, X.P., Xie, H.X., Ao, W.Q., Li, J.Q., Acta Physica Sinica, 59[5] 2010, 3350-3356.

[190] Forster, P.M., Sleight, A.W., International Journal of Inorganic Materials, 1[2] 1999, 123-127. https://doi.org/10.1016/S1466-6049(99)00021-5

[191] Zhang, Z., Yang, L., Liu, H., Pan, K., Wang, W., Zeng, X., Chen, X., Ceramics International, 42[16] 2016, 18902-18906. https://doi.org/10.1016/j.ceramint.2016.09.039

[192] Das, S., Das, S., Das, K., Journal of Materials Engineering and Performance, 22[11] 2013, 3357-3363. https://doi.org/10.1007/s11665-013-0652-6

[193] Yuan, C., Liang, Y., Wang, J., Liang, E., Journal of the Chinese Ceramic Society, 37[5] 2009, 728-732.

[194] Sumithra, S., Tyagi, A.K., Umarji, A.M., Materials Science and Engineering B, 116[1] 2005, 14-18. https://doi.org/10.1016/j.mseb.2004.08.015

[195] Woodcock, D.A., Lightfoot, P., Ritter, C., Journal of Solid State Chemistry, 149[1] 2000, 92-98. https://doi.org/10.1006/jssc.1999.8502

[196] Rimmer, L.H.N., Dove, M.T., Journal of Physics - Condensed Matter, 27[18] 2015, 185401. https://doi.org/10.1088/0953-8984/27/18/185401

[197] Sumithra, S., Waghmare, U.V., Umarji, A.M., Physical Review B, 76[2] 2007, 024307. https://doi.org/10.1103/PhysRevB.76.024307

[198] Liu, H., Zhang, Z., Zhang, W., Zeng, X., Chen, X., Ceramics International, 39[3] 2013, 2781-2786. https://doi.org/10.1016/j.ceramint.2012.09.045

[199] Peng, J., Wu, M.M., Wang, H., Hao, Y.M., Hu, Z., Yu, Z.X., Chen, D.F., Kiyanagi, R., Fieramosca, J.S., Short, S., Jorgensen, J., Journal of Alloys and Compounds, 453[1-2] 2008, 49-54. https://doi.org/10.1016/j.jallcom.2006.11.208

[200] Peng, J., Wu, M.M., Guo, F.L., Han, S.B., Liu, Y.T., Chen, D.F., Zhao, X.H., Hu, Z., Journal of Materials Science, 46[15] 2011, 5160-5164. https://doi.org/10.1007/s10853-011-5447-2

[201] Sumithra, S., Tyagi, A.K., Umarji, A.M., Materials Science and Engineering B, 116[1] 2005, 14-18. https://doi.org/10.1016/j.mseb.2004.08.015

[202] Peng, J., Liu, X.Z., Guo, F.L., Han, S.B., Liu, Y.T., Chen, D.F., Hu, Z., Materials at High Temperatures, 27[2] 2010, 151-156. https://doi.org/10.3184/096034010X12743509428372

[203] Liu, H., Zhang, W., Zhang, Z., Chen, X., Ceramics International, 38[4] 2012, 2951-2956. https://doi.org/10.1016/j.ceramint.2011.11.072

[204] Mary, T.A., Evans, J.S.O., Vogt, T., Sleight, A.W., Science, 272[5258], 1996, 90-92. https://doi.org/10.1126/science.272.5258.90

[205] Cao, D., Bridges, F., Kowach, G.R., Ramirez, A.P., Physical Review B, 68[1] 2003, 143031-1430314. https://doi.org/10.1103/PhysRevB.68.014303

[206] Evans, J.S.O., Mary, T.A., Vogt, T., Subramanian, M.A., Sleight, A.W., Chemistry of Materials, 8[12] 1996, 2809-2823. https://doi.org/10.1021/cm9602959

[207] Pryde, A.K.A., Hammonds, K.D., Dove, M.T., Heine, V., Gale, J.D., Warren, M.C., Phase Transitions, 61[1-4B] 1997, 141-153.

[208] Evans, J.S.O., Mary, T.A., Sleight, A.W., Physica B, 241-243, 1997, 311-316. https://doi.org/10.1016/S0921-4526(97)00571-1

[209] Pryde, A.K.A., Hammonds, K.D., Dove, M.T., Heine, V., Gale, J.D., Warren, M.C., Journal of Physics Condensed Matter, 8[50] 1996, 10973-10982. https://doi.org/10.1088/0953-8984/8/50/023

[210] Mittal, R., Chaplot, S.L., Physical Review B, 60[10] 1999, 7234-7237. https://doi.org/10.1103/PhysRevB.60.7234

[211] Cao, D., Bridges, F., Kowach, G.R., Ramirez, A.P., Physical Review Letters, 89[21] 2002, 2159021-2159024. https://doi.org/10.1103/PhysRevLett.89.215902

[212] Boerio-Goates, J., Stevens, R., Lang, B., Woodfield, B.F., Journal of Thermal Analysis and Calorimetry, 69[3] 2002, 773-783. https://doi.org/10.1023/A:1020639502445

[213] Stevens, R., Linford, J., Woodfield, B.F., Boerio-Goates, J., Lind, C., Wilkinson, A.P., Kowach, G., Journal of Chemical Thermodynamics, 35[6] 2003, 919-937. https://doi.org/10.1016/S0021-9614(03)00050-8

[214] Yamamura, Y., Tsuji, T., Saito, K., Sorai, M., Journal of Chemical Thermodynamics, 36[6] 2004, 525-531. https://doi.org/10.1016/j.jct.2004.03.009

[215] Ravindran, T.R., Arora, A.K., Mary, T.A., Physical Review B, 67[6] 2003, 643011-643014. https://doi.org/10.1103/PhysRevB.67.064301

[216] Schlesinger, Z., Rosen, J.A., Hancock, J.N., Ramirez, A.P., Physical Review Letters, 101[1] 2008, 015501. https://doi.org/10.1103/PhysRevLett.101.015501

[217] Gupta, M.K., Mittal, R., Chaplot, S.L., Physical Review B, 88[1] 2013, 014303. https://doi.org/10.1103/PhysRevB.88.014303

[218] Mittal, R., Chaplot, S.L., Pintschovius, L., Achary, S.N., Kowach, G.R., Journal of Physics: Conference Series, 92[1] 2007, 012174. https://doi.org/10.1088/1742-6596/92/1/012174

[219] Govindaraj, R., Sundar, C.S., Arora, A.K., Physical Review B, 76[1] 2007, 012104. https://doi.org/10.1103/PhysRevB.76.012104

[220] Tucker, M.G., Goodwin, A.L., Dove, M.T., Keen, D.A., Wells, S.A., Evans, J.S.O., Physical Review Letters, 95[25] 2005, 255501. https://doi.org/10.1103/PhysRevLett.95.255501

[221] Sanson, A., Chemistry of Materials, 26[12] 2014, 3716-3720. https://doi.org/10.1021/cm501107w

[222] Bridges, F., Keiber, T., Juhas, P., Billinge, S.J.L., Sutton, L., Wilde, J., Kowach, G.R., Physical Review Letters, 112[4] 2014, 045505. https://doi.org/10.1103/PhysRevLett.112.045505

[223] Liang, E.J., Liang, Y., Zhao, Y., Liu, J., Jiang, Y., Journal of Physical Chemistry A, 112[49] 2008, 12582-12587. https://doi.org/10.1021/jp805256d

[224] Hancock, J.N., Turpen, C., Schlesinger, Z., Kowach, G.R., Ramirez, A.P., Physical Review Letters, 93[22] 2004, 225501. https://doi.org/10.1103/PhysRevLett.93.225501

[225] White, M.A., Whitman, C.A., Materials Research Society Symposium Proceedings, 1044, 2008, 293-298.

[226] Kennedy, C.A., White, M.A., Solid State Communications, 134[4] 2005, 271-276. https://doi.org/10.1016/j.ssc.2005.01.031

[227] Kojima, A., Kuroiwa, Y., Aoyagi, S., Sawada, A., Yamamura, Y., Nakajima, N., Tsuji, T., Journal of the Korean Physical Society, 42S, 2003, S1257-S1260.

[228] Ramirez, A.P., Kowach, G.R., Physical Review Letters, 80[22] 1998, 4903-4910.

https://doi.org/10.1103/PhysRevLett.80.4903

[229] David, W.I.F., Evans, J.S.O., Sleight, A.W., Europhysics Letters, 46[5] 1999, 661-666. https://doi.org/10.1209/epl/i1999-00316-7

[230] Mittal, R., Chaplot, S.L., Schober, H., Mary, T.A., Physical Review Letters, 86[20] 2001, 4692-4695. https://doi.org/10.1103/PhysRevLett.86.4692

[231] Wang, K., Reeber, R.R., Applied Physics Letters, 76[16] 2000, 2203-2204. https://doi.org/10.1063/1.126296

[232] Sikka, S.K., Journal of Physics Condensed Matter, 16[14] 2004, S1033-S1039. https://doi.org/10.1088/0953-8984/16/14/013

[233] Gava, V., Martinotto, A.L., Perottoni, C.A., Physical Review Letters, 109[19] 2012, 195503 https://doi.org/10.1103/PhysRevLett.109.195503

[234] Perottoni, C.A., Da Jornada, J.A.H., Science, 280[5365] 1998, 886-889. https://doi.org/10.1126/science.280.5365.886

[235] Drymiotis, F.R., Ledbetter, H., Betts, J.B., Kimura, T., Lashley, J.C., Migliori, A., Ramirez, A., Kowach, G., Van Duijn, J., Physical Review Letters, 93[2] 2004, 025502-1. https://doi.org/10.1103/PhysRevLett.93.025502

[236] Ulbrich, N., Tröger, W., Butz, T., Blaha, P., Zeitschrift fur Naturforschung A, 55[1-2] 2000, 301-310. https://doi.org/10.1515/zna-2000-1-253

[237] Evans, J.S.O., David, W.I.F., Sleight, A.W., Acta Crystallographica B, 55[3] 1999, 333-340. https://doi.org/10.1107/S0108768198016966

[238] Hashimoto, T., Morito, Y., Journal of the Ceramic Society of Japan, 110[1285] 2002, 823-825. https://doi.org/10.2109/jcersj.110.823

[239] De Meyer, C., Van Driessche, I., Hoste, S., Key Engineering Materials, 206-213[1] 2001, 11-14.

[240] Chen, J.C., Huang, G.C., Hu, C., Weng, J.P., Scripta Materialia, 49[3] 2003, 261-266. https://doi.org/10.1016/S1359-6462(03)00213-6

[241] Reznitskii, L.A., Inorganic Materials, 38[10] 2002, 1020-1021. https://doi.org/10.1023/A:1020525321044

[242] Liu, H., Cheng, X., Zhang, Z., Journal of Jiangsu University - Natural Science, 30[3] 2009, 265-269.

[243] Cheng, X.N., Sun, X.J., Yang, J., Xu, G.F., Journal of Jiangsu University - Natural Science, 26[4] 2005, 350-353.

[244] Du, X., Qin, M., Guo, R., Qu, X., Powder Metallurgy Technology, 26[1] 2008, 57-59+64.

[245] Sun, X.J., Yang, J., Liu, Q.Q., Cheng, X.N., Chinese Journal of Inorganic Chemistry, 21[9] 2005, 1412-1416.

[246] Liu, Q.Q., Yang, J., Sun, X.J., Yan, X.H., Cheng, X.N., Journal of Materials Engineering, S1, 2006, 147-150.

[247] Sun, X.J., Yang, J., Liu, Q.Q., Cheng, X.N., Chinese Journal of Inorganic Chemistry, 22[9] 2006, 1635-1639.

[248] Du, X., Qu, X., Qin, M., Guo, R., Journal of University of Science and Technology, Beijing, 29[9] 2007, 925-927.

[249] Yang, J., Liu, Q., Zang, C., Cheng, X., Advanced Materials Research, 177, 2011, 245-248. https://doi.org/10.4028/www.scientific.net/AMR.177.245

[250] Sun, X., Yang, J., Liu, Q., Cheng, X., Journal of Alloys and Compounds, 481[1-2] 2009, 668-672. https://doi.org/10.1016/j.jallcom.2009.03.061

[251] Kozy, L.C., Tahir, M.N., Lind, C., Tremel, W., Journal of Materials Chemistry, 19[18] 2009, 2760-2765. https://doi.org/10.1039/b820014a

[252] Morito, Y., Wang, S., Ohshima, Y., Uehara, T., Hashimoto, T., Journal of the Ceramic Society of Japan, 110[1282] 2002, 544-548. https://doi.org/10.2109/jcersj.110.544

[253] Nishiyama, S., Hayashi, T., Hattori, T., Journal of Alloys and Compounds, 417[1-2] 2006, 187-189. https://doi.org/10.1016/j.jallcom.2005.07.075

[254] Badrinarayanan, P., Ahmad, I., Akinc, M., Kessler, M.R., Materials Chemistry and Physics, 131[1-2] 2011, 12-17. https://doi.org/10.1016/j.matchemphys.2011.09.055

[255] Sun, X., Yang, J., Liu, Q., Cheng, X., Journal of the Chinese Ceramic Society, 36[1] 2008, 35-39.

[256] Xing, Q., Xing, X., Du, L., Yu, R., Chen, J., Deng, J., Luo, J., Acta Metallurgica Sinica, 41[6], 2005 669-672.

[257] Sun, X.J., Cheng, X.N., Yang, J., Liu, Q.Q., Journal of Functional Materials, 46[8] 2015, 08023-08027.

[258] Sun, X., Zuo, F., Jiao, L., Cheng, X., Yang, J., Liu, Q., Chen, T., Journal of Functional Materials, 47[12] 2016, 12110-12113.

[259] Sun, X.J., Yang, J., Liu, Q.Q., Cheng, X.N., Journal of Functional Materials, 37[11] 2006, 1768-1770+1777.

[260] Sun, X., Cheng, X., Yang, J., Liu, Q., Ceramics International, 39[1] 2013, 165-170. https://doi.org/10.1016/j.ceramint.2012.06.005

[261] Liu, H., Zhang, Z., Zhang, W., Chen, X., Ceramics International, 38[2] 2012, 1341-1345. https://doi.org/10.1016/j.ceramint.2011.09.010

[262] Yan, X., Yang, X., Cheng, X., Fu, T., Qiu, J., Liu, H., Journal of the Chinese Ceramic Society, 34[9] 2006, 1066-1069+1074.

[263] Liu, H., Zhang, Z., Zhang, W., Chen, X., Cheng, X., Surface and Coatings Technology, 205[21-22] 2011, 5073-5076. https://doi.org/10.1016/j.surfcoat.2011.05.010

[264] Liu, H.F., Cheng, X.N., Zhang, Z.P., Physica Status Solidi B, 245[11] 2008, 2509-2513. https://doi.org/10.1002/pssb.200880261

[265] Liu, H., Cheng, X., Zhang, Z., Fu, T., Chinese Journal of Materials Research, 22[1] 2008, 83-86.

[266] Yan, X.H., Cheng, X.N., Xiao, Z.J., Xu, G.F., Journal of Jiangsu University - Natural Science, 28[2] 2007, 135-137+179.

[267] Sutton, M.S., Talghader, J., Journal of Microelectromechanical Systems, 13[4] 2004, 688-695. https://doi.org/10.1109/JMEMS.2004.832191

[268] Shen, S., Qin, X., Xiang, S., Hao, F., Tan, J., Xiao, Q., Materials Science Forum, 689, 2011, 407-412. https://doi.org/10.4028/www.scientific.net/MSF.689.407

[269] Tsuji, T., Yamamura, Y., Nakajima, N., Thermochimica Acta, 416[1-2] 2004, 93-98. https://doi.org/10.1016/j.tca.2003.04.001

[270] Cao, W., Huang, Q., Rong, Y., Wang, Y., Deng, J., Chen, J., Xing, X., Inorganic Chemistry Frontiers, 3[6] 2016, 856-860. https://doi.org/10.1039/C5QI00292C

[271] Liu, Q., Yang, J., Sun, X., Cheng, X., Physica Status Solidi B, 245[11] 2008, 2477-2482. https://doi.org/10.1002/pssb.200880260

[272] Zhao, X.H., Huang, L., Liu, P.H., Ma, H., Chinese Journal of Chemistry, 21[12] 2003, 1529-1531. https://doi.org/10.1002/cjoc.20030211202

[273] Yang, J., Liu, Q., Sun, X., Xu, G., Cheng, X., Ceramics International, 35[1] 2009, 441-445. https://doi.org/10.1016/j.ceramint.2007.12.006

[274] Zhao, R., Yang, X., Wang, H., Han, J., Ma, H., Zhao, X., Journal of Solid State

Chemistry, 180[11] 2007, 3160-3165. https://doi.org/10.1016/j.jssc.2007.09.011

[275] Liu, Q.Q., Yang, J., Sun, X.J., Cheng, X.N., Journal of Inorganic Materials, 22[1] 2007, 70-74.

[276] Liu, Q.Q., Yang, J., Sun, X.J., Cheng, X.N., Chemical Journal of Chinese Universities, 28[3] 2007, 397-401.

[277] Evans, J.S.O., Hanson, P.A., Ibberson, R.M., Duan, N., Kameswari, U., Sleight, A.W., Journal of the American Chemical Society, 122[36] 2000, 8694-8699. https://doi.org/10.1021/ja0013428

[278] Liu, H., Pan, K., Jin, Q., Zhang, Z., Wang, G., Zeng, X., Ceramics International, 40[3] 2014, 3873-3878. https://doi.org/10.1016/j.ceramint.2013.08.028

[279] Liu, H., Ma, J., Zhang, Z., Pan, K., Zhu, J., Zeng, X., Journal of Vacuum Science and Technology, 35[4] 2015, 439-443.

[280] De Buysser, K., Van Driessche, I., Putte, B.V., Schaubroeck, J., Hoste, S., Journal of Solid State Chemistry, 180[8] 2007, 2310-2315. https://doi.org/10.1016/j.jssc.2007.05.028

[281] De Buysser, K., Van Driessche, I., Vande Putte, B., Vanhee, P., Schaubroeck, J., Hoste, S., Inorganic Chemistry, 47[2] 2008, 736-741. https://doi.org/10.1021/ic701660w

[282] Chen, X., Guo, F., Deng, X., Tao, J., Ma, H., Zhao, X., Journal of Alloys and Compounds, 537, 2012, 227-231. https://doi.org/10.1016/j.jallcom.2012.04.106

[283] Nakajima, N., Yamamura, Y., Tsuji, T., Solid State Communications, 128[5] 2003, 193-196. https://doi.org/10.1016/S0038-1098(03)00525-8

[284] Hashimoto, T., Kuwahara, J., Yoshida, T., Nashimoto, M., Takahashi, Y., Takahashi, K., Morito, Y., Solid State Communications, 131[3-4] 2004, 217-221. https://doi.org/10.1016/j.ssc.2004.05.001

[285] Yamamura, Y., Saito, K., Journal of the Physical Society of Japan, 76[12] 2007, 123603. https://doi.org/10.1143/JPSJ.76.123603

[286] Morito, Y., Takahashi, K., Wang, S., Abe, H., Katoh, A., Hashimoto, T., Journal of the Ceramic Society of Japan, 110[1285] 2002, 807-812. https://doi.org/10.2109/jcersj.110.807

[287] Li, H.H., Han, J.S., Ma, H., Huang, L., Zhao, X.H., Journal of Solid State Chemistry, 180[3] 2007, 852-857. https://doi.org/10.1016/j.jssc.2006.12.008

[288] Jakubinek, M.B., Whitman, C.A., White, M.A., Journal of Thermal Analysis and

Calorimetry, 99[1] 2010, 165-172. https://doi.org/10.1007/s10973-009-0458-9

[289] Yang, J., Yang, Y., Liu, Q., Xu, G., Cheng, X., Journal of Materials Science and Technology, 26[7] 2010, 665-668. https://doi.org/10.1016/S1005-0302(10)60103-X

[290] Yamashina, N., Isobe, T., Ando, S., Journal of Photopolymer Science and Technology, 25[3], 2012, 385-388. https://doi.org/10.2494/photopolymer.25.385

[291] Sharma, G.R., Lind, C., Coleman, M.R., Materials Chemistry and Physics, 137[2] 2012, 448-457. https://doi.org/10.1016/j.matchemphys.2012.09.009

[292] Neely, L.A., Kochergin, V., See, E.M., Robinson, H.D., Journal of Materials Science, 49[1] 2014, 392-396. https://doi.org/10.1007/s10853-013-7716-8

[293] Isobe, T., Kato, Y., Mizutani, M., Ota, T., Daimon, K., Materials Letters, 62[23] 2008, 3913-3915. https://doi.org/10.1016/j.matlet.2008.05.046

[294] Balch, D.K., Dunand, D.C., Metallurgical and Materials Transactions A, 35[13] 2004, 1159-1165. https://doi.org/10.1007/s11661-004-1019-2

[295] Dedova, E.S., Pertushina, M.U., Kondratenko, A.I., Gorev, M.V., Kulkov, S.N., AIP Conference Proceedings, 1783, 2016, 020037.

[296] Chang, S.Y., Tsai, H.H., Wang, S.C., Applied Mechanics and Materials, 44-47, 2011, 2950-2953. https://doi.org/10.4028/www.scientific.net/AMM.44-47.2950

[297] Hancock, J.C., Chapman, K.W., Halder, G.J., Morelock, C.R., Kaplan, B.S., Gallington, L.C., Bongiorno, A., Han, C., Zhou, S., Wilkinson, A.P., Chemistry of Materials, 27[11] 2015, 3912-3918. https://doi.org/10.1021/acs.chemmater.5b00662

[298] Hester, B.R., Hancock, J.C., Lapidus, S.H., Wilkinson, A.P., Chemistry of Materials, 29[2] 2017, 823-831. https://doi.org/10.1021/acs.chemmater.6b04809

[299] Hu, L., Chen, J., Fan, L., Deng, J., Yu, R., Xing, X., Journal of the American Ceramic Society, 97[4] 2014, 1009-1011. https://doi.org/10.1111/jace.12855

[300] Greve, B.K., Martin, K.L., Lee, P.L., Chupas, P.J., Chapman, K.W., Wilkinson, A.P., Journal of the American Chemical Society, 132[44] 2010, 15496-15498. https://doi.org/10.1021/ja106711v

[301] Liu, Y., Wang, Z., Wu, M., Sun, Q., Chao, M., Jia, Y., Computational Materials Science, 107, 2015, 157-162. https://doi.org/10.1016/j.commatsci.2015.05.019

[302] Lazar, P., Bučko, T., Hafner, J., Physical Review B, 92[22] 2015, 224301. https://doi.org/10.1103/PhysRevB.92.224301

[303] Li, C.W., Tang, X., Mu-oz, J.A., Keith, J.B., Tracy, S.J., Abernathy, D.L., Fultz, B., Physical Review Letters, 107[19] 2011, 195504. https://doi.org/10.1103/PhysRevLett.107.195504

[304] Purans, J., Piskunov, S., Bocharov, D., Kalinko, A., Kuzmin, A., Ali, S.E., Rocca, F., Journal of Physics: Conference Series, 712[1] 2016, 012013. https://doi.org/10.1088/1742-6596/712/1/012013

[305] Van Roekeghem, A., Carrete, J., Mingo, N., Physical Review B, 94[2] 2016, 020303. https://doi.org/10.1103/PhysRevB.94.020303

[306] Handunkanda, S.U., Occhialini, C.A., Said, A.H., Hancock, J.N., Physical Review B, 94[21] 2016, 214102. https://doi.org/10.1103/PhysRevB.94.214102

[307] Hu, L., Chen, J., Sanson, A., Wu, H., Guglieri Rodriguez, C., Olivi, L., Ren, Y., Fan, L., Deng, J., Xing, X., Journal of the American Chemical Society, 138[27] 2016, 8320-8323. https://doi.org/10.1021/jacs.6b02370

[308] Handunkanda, S.U., Curry, E.B., Voronov, V., Said, A.H., Guzmán-Verri, G.G., Brierley, R.T., Littlewood, P.B., Hancock, J.N., Physical Review B, 92[13] 2015, 134101. https://doi.org/10.1103/PhysRevB.92.134101

[309] Yang, C., Tong, P., Lin, J.C., Guo, X.G., Zhang, K., Wang, M., Wu, Y., Lin, S., Huang, P.C., Xu, W., Song, W.H., Sun, Y.P., Applied Physics Letters, 109[2] 2016, 023110. https://doi.org/10.1063/1.4959083

[310] Morelock, C.R., Greve, B.K., Gallington, L.C., Chapman, K.W., Wilkinson, A.P., Journal of Applied Physics, 114[21] 2013, 213501. https://doi.org/10.1063/1.4836855

[311] Morelock, C.R., Gallington, L.C., Wilkinson, A.P., Chemistry of Materials, 26[5] 2014, 1936-1940. https://doi.org/10.1021/cm5002048

[312] Wang, L., Wang, C., Sun, Y., Deng, S., Shi, K., Lu, H., Hu, P., Zhang, X., Journal of the American Ceramic Society, 98[9] 2015, 2852-2857. https://doi.org/10.1111/jace.13676

[313] Wang, L., Yuan, P.F., Wang, F., Sun, Q., Liang, E.J., Jia, Y., Guo, Z.X., Physics Letters A, 378[38-39] 2014, 2906-2909. https://doi.org/10.1016/j.physleta.2014.08.001

[314] Chatterji, T., Zbiri, M., Hansen, T.C., Applied Physics Letters, 98[18] 2011, 181911. https://doi.org/10.1063/1.3588414

[315] Wang, L., Yuan, P.F., Wang, F., Sun, Q., Liang, E.J., Jia, Y., Materials Research

Bulletin, 47[5] 2012, 1113-1118.
https://doi.org/10.1016/j.materresbull.2012.02.020

[316] Woodcock, D.A., Lightfoot, P., Villaescusa, L.A., Díaz-Caba-as, M.J., Camblor, M.A., Engberg, D., Chemistry of Materials, 11[9] 1999, 2508-2514.
https://doi.org/10.1021/cm991047q

[317] Woodcock, D.A., Lightfoot, P., Wright, P.A., Villaescusa, L.A., Díaz-Caba-as, M.J., Camblor, M.A., Journal of Materials Chemistry, 9[2] 1999, 349-351.
https://doi.org/10.1039/a808059c

[318] Lightfoot, P., Woodcock, D.A., Maple, M.J., Villaescusa, L.A., Wright, P.A., Journal of Materials Chemistry, 11[1] 2001, 212-216.
https://doi.org/10.1039/b002950p

[319] Carey, T., Corma, A., Rey, F., Tang, C.C., Hriljac, J.A., Anderson, P.A., Chemical Communications, 48[44] 2012, 5829-5831. https://doi.org/10.1039/c2cc30582h

[320] Martínez-I-esta, M.M., Lobo, R.F., Journal of Physical Chemistry B, 109[19] 2005, 9389-9396.

[321] Attfield, M.P., Sleight, A.W., Chemistry of Materials, 10[7] 1998, 2013-2019.
https://doi.org/10.1021/cm9801587

[322] Tao, J.Z., Sleight, A.W., Journal of Physics and Chemistry of Solids, 64[9-10] 2003, 1473-1479. https://doi.org/10.1016/S0022-3697(03)00102-1

[323] Amri, M., Walton, R.I., Chemistry of Materials, 21[14] 2009, 3380-3390.
https://doi.org/10.1021/cm901140u

[324] Marinkovic, B.A., Jardim, P.M., Saavedra, A., Lau, L.Y., Baehtz, C., de Avillez, R.R., Rizzo, F., Microporous and Mesoporous Materials, 71[1-3] 2004, 117-124.
https://doi.org/10.1016/j.micromeso.2004.03.023

[325] Villaescusa, L.A., Lightfoot, P., Teat, S.J., Morris, R.E., Journal of the American Chemical Society, 123[23] 2001, 5453-5459. https://doi.org/10.1021/ja015797o

[326] Reisner, B.A., Lee, Y., Hanson, J.C., Jones, G.A., Parise, J.B., Corbin, D.R., Toby, B.H., Freitag, A., Larese, J.Z., Kahlenberg, V., Chemical Communications, 22, 2000, 2221-2222. https://doi.org/10.1039/b006929i

[327] Attfield, M.P., Chemical Communications, 5, 1998, 601-602.
https://doi.org/10.1039/a707141h

[328] Attfield, M.P., Feygenson, M., Neuefeind, J.C., Proffen, T.E., Lucas, T.C.A., Hriljac, J.A., RSC Advances, 6[24] 2016, 19903-19909.

https://doi.org/10.1039/C5RA23827G

[329] Leardini, L., Quartieri, S., Vezzalini, G., Arletti, R., Microporous and Mesoporous Materials, 202[C] 2015, 226-233. https://doi.org/10.1016/j.micromeso.2014.10.006

[330] Bhange, D.S., Ramaswamy, V., Materials Research Bulletin, 41[7] 2006, 1392-1402. https://doi.org/10.1016/j.materresbull.2005.12.002

[331] Bhange, D.S., Ramaswamy, V., Microporous and Mesoporous Materials, 130[1-3] 2010, 322-326. https://doi.org/10.1016/j.micromeso.2009.11.029

[332] Goodwin, A.L., Chapman, K.W., Kepert, C.J., Journal of the American Chemical Society, 127[51] 2005, 17980-17981. https://doi.org/10.1021/ja056460f

[333] Goodwin, A.L., Kepert, C.J., Physical Review B, 71[14] 2005, 140301. https://doi.org/10.1103/PhysRevB.71.140301

[334] Hibble, S.J., Wood, G.B., Bilbé, E.J., Pohl, A.H., Tucker, M.G., Hannon, A.C., Chippindale, A.M., Zeitschrift fur Kristallographie, 225[11] 2010, 457-462. https://doi.org/10.1524/zkri.2010.1314

[335] Goodwin, A.L., Calleja, M., Conterio, M.J., Dove, M.T., Evans, J.S.O., Keen, D.A., Peters, L., Tucker, M.G., Science, 319[5864] 2008, 794-797. https://doi.org/10.1126/science.1151442

[336] Calleja, M., Goodwin, A.L., Dove, M.T., Journal of Physics - Condensed Matter, 20[25] 2008, 255226. https://doi.org/10.1088/0953-8984/20/25/255226

[337] Ovens, J.S., Leznoff, D.B., Inorganic Chemistry, 56[13] 2017, 7332-7343. https://doi.org/10.1021/acs.inorgchem.6b03153

[338] Chapman, K.W., Chupas, P.J., Kepert, C.J., Journal of the American Chemical Society, 128[21] 2006, 7009-7014. https://doi.org/10.1021/ja060916r

[339] Phillips, A.E., Goodwin, A.L., Halder, G.J., Southon, P.D., Kepert, C.J., Angewandte Chemie, 47[8] 2008, 1396-1399 https://doi.org/10.1002/anie.200704421

[340] Korčok, J.L., Katz, M.J., Leznoff, D.B., Journal of the American Chemical Society, 131[13] 2009, 4866-4871. https://doi.org/10.1021/ja809631r

[341] Reczyński, M., Chorazy, S., Nowicka, B., Sieklucka, B., Ohkoshi, S.I., Inorganic Chemistry, 56[1] 2017, 179-185. https://doi.org/10.1021/acs.inorgchem.6b01883

[342] Kroll, P., Andrade, M., Yan, X., Ionescu, E., Miehe, G., Riedel, R., Journal of Physical Chemistry C, 116[1] 2012, 526-531. https://doi.org/10.1021/jp2106583

[343] Chapman, K.W., Chupas, P.J., Kepert, C.J., Journal of the American Chemical Society, 127[44] 2005, 15630-15636. https://doi.org/10.1021/ja055197f

[344] Zwanziger, J.W., Physical Review B, 76[5] 2007, 052102. https://doi.org/10.1103/PhysRevB.76.052102

[345] Ravindran, T.R., Arora, A.K., Chandra, S., Valsakumar, M.C., Chandra Shekar, N.V., Physical Review B, 76[5] 2007, 054302. https://doi.org/10.1103/PhysRevB.76.054302

[346] Ding, P., Liang, E.J., Jia, Y., Du, Z.Y., Journal of Physics - Condensed Matter, 20[27] 2008, 275224. https://doi.org/10.1088/0953-8984/20/27/275224

[347] Mittal, R., Chaplot, S.L., Schober, H., Applied Physics Letters, 95[20] 2009, 201901. https://doi.org/10.1063/1.3264963

[348] Chapman, K.W., Chupas, P.J., Journal of the American Chemical Society, 129[33] 2007, 10090-10091. https://doi.org/10.1021/ja073791e

[349] Fang, H., Dove, M.T., Rimmer, L.H.N., Misquitta, A.J., Physical Review B, 88[10] 2013, 104306. https://doi.org/10.1103/PhysRevB.88.104306

[350] Hibble, S.J., Chippindale, A.M., Marelli, E., Kroeker, S., Michaelis, V.K., Greer, B.J., Aguiar, P.M., Bilbé, E.J., Barney, E.R., Hannon, A.C., Journal of the American Chemical Society, 135[44] 2013, 16478-16489. https://doi.org/10.1021/ja406848s

[351] Fairbank, V.E., Thompson, A.L., Cooper, R.I., Goodwin, A.L., Physical Review B, 86[10] 2012, 104113. https://doi.org/10.1103/PhysRevB.86.104113

[352] Adak, S., Daemen, L.L., Nakotte, H., Journal of Physics - Conference Series, 251[1] 2010, 012007. https://doi.org/10.1088/1742-6596/251/1/012007

[353] Strzhemechny, M.A., Legchenkova, I.V., Low Temperature Physics, 36[5] 2010, 370-372. https://doi.org/10.1063/1.3432243

[354] Brown, S., Cao, J., Musfeldt, J.L., Dragoe, N., Cimpoesu, F., Ito, S., Takagi, H., Cross, R.J., Physical Review B, 73[12] 2006, 12544. https://doi.org/10.1103/PhysRevB.73.125446

[355] Aleksandrovskii, A.N., Eselson, V.B., Manzhelii, V.G., Udovidchenko, B.G., Soldatov, A.V., Sundqvist, B., Low Temperature Physics, 23[11] 1997, 943-946. https://doi.org/10.1063/1.593498

[356] Liu, Y., Jia, Y., Sun, Q., Liang, E., Physics Letters A, 379[1-2] 2015, 54-59. https://doi.org/10.1016/j.physleta.2014.10.041

[357] Margadonna, S., Arvanitidis, J., Papagelis, K., Prassides, K., Chemistry of Materials, 17[17] 2005, 4474-4478. https://doi.org/10.1021/cm051341k

[358] Su, Y., Wei, H., Gao, R., Yang, Z., Zhang, J., Zhong, Z., Zhang, Y., Carbon, 50[8] 2012, 2804-2809. https://doi.org/10.1016/j.carbon.2012.02.045

[359] Hu, Y., Chen, J., Wang, B., Carbon, 95, 2015, 239-249. https://doi.org/10.1016/j.carbon.2015.08.022

[360] Yoon, D., Son, Y.W., Cheong, H., Nano Letters, 11[8] 2011, 3227-3231. https://doi.org/10.1021/nl201488g

[361] Mann, S., Kumar, R., Jindal, V.K., RSC Advances, 7[36] 2017, 22378-22387. https://doi.org/10.1039/C7RA01591G

[362] Song, X., Sun, Z., Acta Metallurgica Sinica, 47[11] 2011, 1362-1371.

[363] Qu, B., He, H., Pan, B., AIP Advances, 6[7] 2016, 075122. https://doi.org/10.1063/1.4960428

[364] Ichigo, M., Takenaka, K., Journal of the Japan Institute of Metals, 77[10] 2013, 415-418. https://doi.org/10.2320/jinstmet.JAW201302

[365] Takenaka, K., Ichigo, M., Composite Science and Technology, 104, 2014, 47-51. https://doi.org/10.1016/j.compscitech.2014.08.029

[366] Takenaka, K., Kuzuoka, K., Sugimoto, N., Journal of Applied Physics, 118[8] 2015, 084902. https://doi.org/10.1063/1.4929363

[367] Deng, S., Sun, Y., Yan, J., Shi, Z., Shi, K., Wang, L., Hu, P., Malik, M.I., Wang, C., Solid State Communications, 222, 2015, 37-41. https://doi.org/10.1016/j.ssc.2015.08.024

[368] Yan, X., Miao, J., Liu, J., Wu, X., Zou, H., Sha, D., Cheng, X., Advances in Applied Ceramics, 115[7] 2016, 422-426. https://doi.org/10.1080/17436753.2016.1172412

[369] Guo, X.G., Tong, P., Lin, J.C., Yang, C., Zhang, K., Wang, M., Wu, Y., Lin, S., Song, W.H., Sun, Y.P., Scripta Materialia, 128, 2017, 74-77. https://doi.org/10.1016/j.scriptamat.2016.10.002

[370] Sun, Y., Wang, C., Wen, Y., Materials Science Forum, 561-565[1] 2007, 557-562. https://doi.org/10.4028/www.scientific.net/MSF.561-565.557

[371] Iikubo, S., Kodama, K., Takenaka, K., Takagi, H., Takigawa, M., Shamoto, S., Physical Review Letters, 101[20] 2008, 205901. https://doi.org/10.1103/PhysRevLett.101.205901

[372] Takenaka, K., Asano, K., Misawa, M., Takagi, H., Applied Physics Letters, 92[1] 2008, 011927. https://doi.org/10.1063/1.2831715

[373] Sun, Z., Song, X., Journal of Materials Science and Technology, 30[9] 2014, 903-909. https://doi.org/10.1016/j.jmst.2013.12.004

[374] Tan, J., Huang, R., Wang, W., Li, W., Zhao, Y., Li, S., Han, Y., Huang, C., Li, L., Nano Research, 8[7] 2015, 2302-2307. https://doi.org/10.1007/s12274-015-0740-z

[375] Sun, Z.H., Song, X.Y., Yin, F.X., Sun, L.X., Yuan, X.K., Liu, X.M., Journal of Physics D, 42[12] 2009, 122004. https://doi.org/10.1088/0022-3727/42/12/122004

[376] Qu, B.Y., He, H.Y., Pan, B.C., Applied Physics A, 114[3] 2014, 785-791. https://doi.org/10.1007/s00339-013-7650-2

[377] Qu, B.Y., He, H.Y., Pan, B.C., Advances in Condensed Matter Physics, 2012, 913168.

[378] Kodama, K., Iikubo, S., Takenaka, K., Takigawa, M., Takagi, H., Shamoto, S., Physical Review B, 81[22] 2010, 224419. https://doi.org/10.1103/PhysRevB.81.224419

[379] Takenaka, K., Takagi, H., Journal of the Japan Institute of Metals, 70[9] 2006, 764-768. https://doi.org/10.2320/jinstmet.70.764

[380] Takenaka, K., Takagi, H., Materials Transactions, 47[3] 2006, 471-474. https://doi.org/10.2320/matertrans.47.471

[381] Nakamura, Y., Takenaka, K., Kishimoto, A., Takagi, H., Journal of the American Ceramic Society, 92[12] 2009, 2999-3003. https://doi.org/10.1111/j.1551-2916.2009.03297.x

[382] Zhang, C., Zhu, J., Zhang, M., Chinese Journal of Rare Metals, 33[5] 2009, 686-690.

[383] Zhang, C., Zhu, J., Zhang, M., Acta Metallurgica Sinica, 45[1] 2009, 97-101.

[384] Huang, R., Li, L., Cai, F., Xu, X., Qian, L., Applied Physics Letters, 93[8] 2008, 081902. https://doi.org/10.1063/1.2970998

[385] Huang, R.J., Xu, W., Xu, X.D., Li, L.F., Pan, X.Q., Evans, D., Materials Letters, 62[16] 2008, 2381-2384. https://doi.org/10.1016/j.matlet.2007.12.003

[386] Takenaka, K., Takagi, H., Applied Physics Letters, 87[26] 2005, 261902 https://doi.org/10.1063/1.2147726

[387] Matsuno, J., Takenaka, K., Takagi, H., Matsumura, D., Nishihata, Y., Mizuki, J.,

Applied Physics Letters, 94[18] 2009, 181904. https://doi.org/10.1063/1.3129169

[388] Huang, R., Chen, Z., Chu, X., Wu, Z., Li, L., Journal of Composite Materials, 45[16] 2011, 1675-1682. https://doi.org/10.1177/0021998310385031

[389] Huang, R., Wu, Z., Yang, H., Chen, Z., Chu, X., Li, L., Cryogenics, 50[11-12] 2010, 750-753. https://doi.org/10.1016/j.cryogenics.2010.09.001

[390] Ding, L., Wang, C., Sun, Y., Colin, C.V., Chu, L., Journal of Applied Physics, 117[21] 2015, 213915. https://doi.org/10.1063/1.4921537

[391] Lin, J.C., Wang, B.S., Lin, S., Tong, P., Lu, W.J., Zhang, L., Song, W.H., Sun, Y.P., Journal of Applied Physics, 111[4] 2012, 043905. https://doi.org/10.1063/1.3684653

[392] Chen, Z., Huang, R., Chu, X., Wu, Z., Liu, Z., Zhou, Y., Li, L., Cryogenics, 52[11] 2012, 629-631. https://doi.org/10.1016/j.cryogenics.2012.08.009

[393] Guo, X., Lin, J., Tong, P., Sun, Y., Materials China, 34[7-8] 2015, 526-530.

[394] Lin, J.C., Tong, P., Zhou, X.J., Lin, H., Ding, Y.W., Bai, Y.X., Chen, L., Guo, X.G., Yang, C., Song, B., Wu, Y., Lin, S., Song, W.H., Sun, Y.P., Applied Physics Letters, 107[13] 2015, 131902. https://doi.org/10.1063/1.4932067

[395] Guo, X.G., Lin, J.C., Tong, P., Wang, M., Wu, Y., Yang, C., Song, B., Lin, S., Song, W.H., Sun, Y.P., Applied Physics Letters, 107[20] 2015, 202406. https://doi.org/10.1063/1.4936239

[396] Sun, Y., Wang, C., Wen, Y., Chu, L., Nie, M., Liu, F., Journal of the American Ceramic Society, 93[3] 2010, 650-653. https://doi.org/10.1111/j.1551-2916.2009.03482.x

[397] Zhang, L., Wang, D., Tan, J., Li, W., Wang, W., Huang, R., Li, L., Rare Metal Materials and Engineering, 43[6] 2014, 1304-1307. https://doi.org/10.1016/S1875-5372(14)60112-0

[398] Li, S., Tan, J., Zhao, Y., Wang, W., Huang, R., Li, L., Materials China, 34[7-8] 2015, 521-525.

[399] Tan, J., Huang, R., Li, W., Han, Y., Li, L., Journal of Alloys and Compounds, 593, 2014, 103-105. https://doi.org/10.1016/j.jallcom.2014.01.027

[400] Yan, X., Miao, J., Liu, J., Wu, X., Zou, H., Sha, D., Cheng, X., Advances in Applied Ceramics, 115[7] 2016, 422-426. https://doi.org/10.1080/17436753.2016.1172412

[401] Shi, K., Sun, Y., Deng, S., Wang, L., Hu, P., Wang, C., Materials China, 34[7-8]

2015, 531-535.

[402] Sun, Y., Wang, C., Wen, Y., Chu, L., Pan, H., Nie, M., Tang, M., Journal of the American Ceramic Society, 93[8] 2010, 2178-2181. https://doi.org/10.1111/j.1551-2916.2010.03711.x

[403] Hamada, T., Takenaka, K., Journal of Applied Physics, 109[7] 2011, 07E309.

[404] Qu, B.Y., Pan, B.C., Journal of Applied Physics, 108[11] 2010, 113920. https://doi.org/10.1063/1.3517824

[405] Zhou, W., Wu, H., Yildirim, T., Simpson, J.R., Walker, A.R.H., Physical Review B, 78[5] 2008, 054114. https://doi.org/10.1103/PhysRevB.78.054114

[406] Lock, N., Christensen, M., Kepert, C.J., Iversen, B.B., Chemical Communications, 49[8] 2013, 789-791. https://doi.org/10.1039/C2CC37415C

[407] Lock, N., Wu, Y., Christensen, M., Cameron, L.J., Peterson, V.K., Bridgeman, A.J., Kepert, C.J., Iversen, B.B., Journal of Physical Chemistry C, 114[39] 2010, 16181-16186. https://doi.org/10.1021/jp103212z

[408] Lock, N., Christensen, M., Wu, Y., Peterson, V.K., Thomsen, M.K., Piltz, R.O., Ramirez-Cuesta, A.J., McIntyre, G.J., Norén, K., Kutteh, R., Kepert, C.J., Kearley, G.J., Iversen, B.B., Dalton Transactions, 42[6] 2013, 1996-2007. https://doi.org/10.1039/C2DT31491F

[409] Rimmer, L.H.N., Dove, M.T., Goodwin, A.L., Palmer, D.C., Physical Chemistry Chemical Physics, 16[39] 2014, 21144-21152. https://doi.org/10.1039/C4CP01701C

[410] Wang, L., Wang, C., Sun, Y., Shi, K., Deng, S., Lu, H., Materials Chemistry and Physics, 175, 2016, 138-145. https://doi.org/10.1016/j.matchemphys.2016.03.003

[411] Han, S.S., Goddard, W.A., Journal of Physical Chemistry C, 111[42] 2007, 15185-15191. https://doi.org/10.1021/jp075389s

[412] Wu, Y., Kobayashi, A., Halder, G.J., Peterson, V.K., Chapman, K.W., Lock, N., Southon, P.D., Kepert, C.J., Angewandte Chemie, 47[46] 2008, 8929-8932. https://doi.org/10.1002/anie.200803925

[413] Wu, Y., Peterson, V.K., Luks, E., Darwish, T.A., Kepert, C.J., Angewandte Chemie, 53[20] 2014, 5175-5178.

[414] Cliffe, M.J., Hill, J.A., Murray, C.A., Coudert, F.X., Goodwin, A.L., Physical Chemistry Chemical Physics, 17[17] 2015, 11586-11592. https://doi.org/10.1039/C5CP01307K

[415] Zhang, Z., Jiang, X., Feng, G., Lin, Z., Hu, B., Li, W., Journal of Solid State Chemistry, 233, 2016, 289-293. https://doi.org/10.1016/j.jssc.2015.10.043

[416] Gao, H., Wei, W., Li, Y., Wu, R., Feng, G., Li, W., Materials, 10[2] 2017, 151. https://doi.org/10.3390/ma10020151

[417] Shang, R., Xu, G.C., Wang, Z.M., Gao, S., Chemistry - A European Journal, 20[4] 2014, 1146-1158. https://doi.org/10.1002/chem.201303425

[418] Murao, T., Physica B+C, 143[1-3] 1986, 273-275. https://doi.org/10.1016/0378-4363(86)90116-6

[419] Murao, T., Physical Review B, 35[12] 1987, 6051-6058. https://doi.org/10.1103/PhysRevB.35.6051

[420] Smith, D., The Journal of Chemical Physics, 79[6] 1983, 2995-3001. https://doi.org/10.1063/1.446129

[421] Panda, M.K., Runčevski, T., Chandra Sahoo, S., Belik, A.A., Nath, N.K., Dinnebier, R.E., Naumov, P., Nature Communications, 5, 2014, 4811. https://doi.org/10.1038/ncomms5811

[422] Smeets, S., Lutz, M., Acta Crystallographica C, 67[2] 2011, m50-m55. https://doi.org/10.1107/S0108270111000023

[423] Kennedy, B.J., Kubota, Y., Kato, K., Solid State Communications, 136[3] 2005, 177-180. https://doi.org/10.1016/j.ssc.2005.05.043

[424] Mittal, R., Chaplot, S.L., Mishra, S.K., Bose, P.P., Physical Review B, 75[17] 2007, 174303. https://doi.org/10.1103/PhysRevB.75.174303

[425] Artioli, G., Dapiaggi, M., Fornasini, P., Sanson, A., Rocca, F., Merli, M., Journal of Physics and Chemistry of Solids, 67[9-10] 2006, 1918-1922. https://doi.org/10.1016/j.jpcs.2006.05.043

[426] Sanson, A., Rocca, F., Dalba, G., Fornasini, P., Grisenti, R., Dapiaggi, M., Artioli, G., Physical Review B, 73[21] 2006, 214305. https://doi.org/10.1103/PhysRevB.73.214305

[427] Fornasini, P., Sanson, A., Vaccari, M., Artioli, G., Dapiaggi, M., Journal of Physics: Conference Series, 92[1] 2007, 012153. https://doi.org/10.1088/1742-6596/92/1/012153

[428] Gupta, M.K., Mittal, R., Rols, S., Chaplot, S.L., AIP Conference Proceedings, 1349[A] 2011, 815-816.

[429] Gupta, M.K., Mittal, R., Rols, S., Chaplot, S.L., Physica B, 407[12] 2012, 2146-

2149. https://doi.org/10.1016/j.physb.2012.02.023

[430] Rimmer, L.H.N., Dove, M.T., Winkler, B., Wilson, D.J., Refson, K., Goodwin, A.L., Physical Review B, 89[21] 2014, 214115. https://doi.org/10.1103/PhysRevB.89.214115

[431] Dapiaggi, M., Kim, H., Božin, E.S., Billinge, S.J.L., Artioli, G., Journal of Physics and Chemistry of Solids, 69[9] 2008, 2182-2186. https://doi.org/10.1016/j.jpcs.2008.03.030

[432] Zheng, X.G., Kubozono, H., Yamada, H., Kato, K., Ishiwata, Y., Xu, C.N., Nature Nanotechnology, 3[12] 2008, 724-726. https://doi.org/10.1038/nnano.2008.309

[433] Negishi, H., Kuroiwa, Y., Akamine, H., Aoyagi, S., Sawada, A., Shobu, T., Negishi, S., Sasaki, M., Solid State Communications, 125[1] 2003, 45-49. https://doi.org/10.1016/S0038-1098(02)00625-7

[434] Choosuwan, H., Guo, R., Bhalla, A.S., Balachandran, U., Journal of Applied Physics, 91[8] 2002, 5051-5054. https://doi.org/10.1063/1.1464232

[435] Rodriguez, E.E., Llobet, A., Proffen, T., Melot, B.C., Seshadri, R., Littlewood, P.B., Cheetham, A.K., Journal of Applied Physics, 105[11] 2009, 114901. https://doi.org/10.1063/1.3120783

[436] Chatterji, T., Henry, P.F., Mittal, R., Chaplot, S.L., Physical Review B, 78[13] 2008, 134105. https://doi.org/10.1103/PhysRevB.78.134105

[437] Chatterji, T., Hansen, T.C., Brunelli, M., Henry, P.F., Applied Physics Letters, 94[24] 2009, 241902. https://doi.org/10.1063/1.3155191

[438] Welche, P.R.L., Heine, V., Dove, M.T., Physics and Chemistry of Minerals, 26[1] 1998, 63-77. https://doi.org/10.1007/s002690050161

[439] Huang, L., Kieffer, J., Physical Review Letters, 95[21] 2005, 215901. https://doi.org/10.1103/PhysRevLett.95.215901

[440] Antao, S.M., Acta Crystallographica Section B, 72[2] 2016, 249-262. https://doi.org/10.1107/S205252061600233X

[441] Sakamoto, A., Matano, T., Takeuchi, H., IEICE Transactions on Electronics, E83-C[9] 2000, 1441-1445.

[442] Bahgat, A.A., Al-Hajry, A., El-Desoky, M.M., Physica Status Solidi A, 203[8] 2006, 1999-2006. https://doi.org/10.1002/pssa.200521339

[443] Wang, Z., Wang, F., Wang, L., Jia, Y., Sun, Q., Journal of Applied Physics, 114[6] 2013, 063508. https://doi.org/10.1063/1.4817902

[444] Sleight, A.W., Materials Research Society Symposium - Proceedings, 755, 2003, 381-391.

[445] Filatov, S.K., Zapiski Vsesoyuznogo Mineralogicheskogo Obshchestva, 111[6] 1982, 674-681.

[446] Li, J., Sleight, A.W., Jones, C.Y., Toby, B.H., Journal of Solid State Chemistry, 178[1] 2005, 285-294. https://doi.org/10.1016/j.jssc.2004.11.017

[447] Kim, I.J., Kim, H.C., Han, I.S., Aneziris, C.G., Key Engineering Materials, 280-283[II] 2005, 1179-1184. https://doi.org/10.4028/www.scientific.net/KEM.280-283.1179

[448] Thieme, C., Rüssel, C., Materials, 9[8] 2016, 1-5. https://doi.org/10.3390/ma9080631

[449] Azuma, M., Oka, K., Nabetani, K., Science and Technology of Advanced Materials, 16[3] 2015, 034904. https://doi.org/10.1088/1468-6996/16/3/034904

[450] Azuma, M., Chen, W.T., Seki, H., Czapski, M., Olga, S., Oka, K., Mizumaki, M., Watanuki, T., Ishimatsu, N., Kawamura, N., Ishiwata, S., Tucker, M.G., Shimakawa, Y., Attfield, J.P., Nature Communications, 2[1] 2011, 347. https://doi.org/10.1038/ncomms1361

[451] Liu, Y., Wang, Z., Chang, D., Sun, Q., Chao, M., Jia, Y., Computational Materials Science, 113, 2016, 198-202. https://doi.org/10.1016/j.commatsci.2015.11.041

[452] Oka, K., Mizumaki, M., Sakaguchi, C., Sinclair, A., Ritter, C., Attfield, J.P., Azuma, M., Physical Review B, 88[1] 2013, 014112. https://doi.org/10.1103/PhysRevB.88.014112

[453] Nabetani, K., Oka, K., Azuma, M., Journal of the Japan Society of Powder and Powder Metallurgy, 61[1] 2014, 35-38. https://doi.org/10.2497/jjspm.61.35

[454] Oka, K., Nabetani, K., Sakaguchi, C., Seki, H., Czapski, M., Shimakawa, Y., Azuma, M., Applied Physics Letters, 103[6] 2013, 061909. https://doi.org/10.1063/1.4817976

[455] Nabetani, K., Muramatsu, Y., Oka, K., Nakano, K., Hojo, H., Mizumaki, M., Agui, A., Higo, Y., Hayashi, N., Takano, M., Azuma, M., Applied Physics Letters, 106[6] 2015, 061912. https://doi.org/10.1063/1.4908258

[456] Nakano, K., Oka, K., Watanuki, T., Mizumaki, M., Machida, A., Agui, A., Kim, H., Komiyama, J., Mizokawa, T., Nishikubo, T., Hattori, Y., Ueda, S., Sakai, Y., Azuma, M., Chemistry of Materials, 28[17] 2016, 6062-6067.

https://doi.org/10.1021/acs.chemmater.6b01160

[457] Pryanichnikov, S.V., Titova, S.G., Kalyuzhnaya, G.A., Gorina, Y.I., Slepukhin, P.A., Journal of Experimental and Theoretical Physics, 107[1] 2008, 69-73. https://doi.org/10.1134/S1063776108070066

[458] Gao, C.Y., Xia, H.R., Xu, J.Q., Zhou, C.L., Zhang, H.J., Wang, J.Y., Materials Letters, 63[1] 2009, 139-141. https://doi.org/10.1016/j.matlet.2008.09.037

[459] Takenaka, K., Okamoto, Y., Shinoda, T., Katayama, N., Sakai, Y., Nature Communications, 8, 2017, 14102. https://doi.org/10.1038/ncomms14102

[460] Senn, M.S., Bombardi, A., Murray, C.A., Vecchini, C., Scherillo, A., Luo, X., Cheong, S.W., Physical Review Letters, 114[3] 2015, 035701. https://doi.org/10.1103/PhysRevLett.114.035701

[461] Qi, T.F., Korneta, O.B., Parkin, S., Hu, J., Cao, G., Physical Review B, 85[16] 2012, 165143. https://doi.org/10.1103/PhysRevB.85.165143

[462] Senn, M.S., Murray, C.A., Luo, X., Wang, L., Huang, F.T., Cheong, S.W., Bombardi, A., Ablitt, C., Mostofi, A.A., Bristowe, N.C., Journal of the American Chemical Society, 138[17] 2016, 5479-5482. https://doi.org/10.1021/jacs.5b13192

[463] Huang, L.F., Lu, X.Z., Rondinelli, J.M., Physical Review Letters, 117[11] 2016, 115901. https://doi.org/10.1103/PhysRevLett.117.115901

[464] Li, J., Yokochi, A., Amos, T.G., Sleight, A.W., Chemistry of Materials, 14[6] 2002, 2602-2606. https://doi.org/10.1021/cm011633v

[465] Ahmed, S.I., Dalba, G., Fornasini, P., Vaccari, M., Rocca, F., Sanson, A., Li, J., Sleight, A.W., Physical Review B, 79[10] 2009, 104302. https://doi.org/10.1103/PhysRevB.79.104302

[466] Fu, L., Chao, M., Chen, H., Liu, X., Liu, Y., Yu, J., Liang, E., Li, Y., Xiao, X., Physics Letters, Section A, 378[28-29] 2014, 1909-1912. https://doi.org/10.1016/j.physleta.2014.04.040

[467] Paul, B., Chatterjee, S., Roy, A., Midya, A., Mandal, P., Grover, V., Tyagi, A.K., Physical Review B, 95[5] 2017, 054103. https://doi.org/10.1103/PhysRevB.95.054103

[468] Li, H., Lv, S., Wang, Z., Xia, Y., Bai, Y., Liu, X., Meng, J., Journal of Applied Physics, 111[10] 2012, 103718. https://doi.org/10.1063/1.4721408

[469] Yamada, I., Marukawa, S., Murakami, M., Mori, S., Applied Physics Letters, 105[23] 2014, 231906. https://doi.org/10.1063/1.4903890

[470] Vyas, J.C., Singh, S.G., AIP Conference Proceedings, 1512, 2013, 912-913. https://doi.org/10.1063/1.4791336

[471] Klimczuk, T., Walker, H.C., Springell, R., Shick, A.B., Hill, A.H., Gaczyński, P., Gofryk, K., Kimber, S.A.J., Ritter, C., Colineau, E., Griveau, J.C., Bouëxière, D., Eloirdi, R., Cava, R.J., Caciuffo, R., Physical Review B, 85[17] 2012, 174506. https://doi.org/10.1103/PhysRevB.85.174506

[472] Singh, S.P., Pandey, D., Yoon, S., Baik, S., Shin, N., Applied Physics Letters, 90[24] 2007, 242915. https://doi.org/10.1063/1.2748856

[473] Wang, F., Xie, Y., Chen, J., Fu, H., Xing, X., Applied Physics Letters, 103[22] 2013, 221901. https://doi.org/10.1063/1.4833280

[474] Fang, H., Wang, Y., Shang, S., Liu, Z.K., Physical Review B, 91[2] 2015, 024104. https://doi.org/10.1103/PhysRevB.91.024104

[475] Jiao, Y.C., Li, M., Qu, B.Y., Wu, M.Y., Zhang, N., Guo, P., Wang, J.J., Computational Materials Science, 124, 2016, 92-97. https://doi.org/10.1016/j.commatsci.2016.07.010

[476] Peng, X., Rong, Y., Fan, L., Lin, K., Zhu, H., Deng, J., Chen, J., Xing, X., Inorganic Chemistry Frontiers, 2[12] 2015, 1091-1094. https://doi.org/10.1039/C5QI00154D

[477] Wang, F., Xie, Y., Chen, J., Fu, H., Xing, X., Physical Chemistry Chemical Physics, 16[11] 2014, 5237-5241. https://doi.org/10.1039/c3cp53197j

[478] Wang, F.F., Cao, Z.M., Chen, J., Xing, X.R., Acta Physico-Chimica Sinica, 30[8] 2014, 1432-1436.

[479] Hu, P., Cao, Z., Chen, J., Deng, J., Sun, C., Yu, R., Xing, X., Materials Letters, 62[30] 2008, 4585-4587. https://doi.org/10.1016/j.matlet.2008.08.028

[480] Sun, C., Cao, Z., Chen, J., Yu, R., Sun, X., Hu, P., Liu, G., Xing, X., Physica Status Solidi B, 245[11] 2008, 2520-2523. https://doi.org/10.1002/pssb.200880265

[481] Chen, J., Wang, F., Huang, Q., Hu, L., Song, X., Deng, J., Yu, R., Xing, X., Scientific Reports, 3, 2013, 2458. https://doi.org/10.1038/srep02458

[482] Chen, J., Nittala, K., Forrester, J.S., Jones, J.L., Deng, J., Yu, R., Xing, X., Journal of the American Chemical Society, 133[29] 2011, 11114-11117. https://doi.org/10.1021/ja2046292

[483] Chen, J., Xing, X.R., Liu, G.R., Li, J.H., Liu, Y.T., Applied Physics Letters, 89[10] 2006, 101914. https://doi.org/10.1063/1.2347279

[484] Peng, X., Chen, J., Lin, K., Fan, L., Rong, Y., Deng, J., Xing, X., RSC Advances, 6[39] 2016, 32979-32982. https://doi.org/10.1039/C6RA03774G

[485] Liu, H., Chen, J., Jiang, X., Pan, Z., Zhang, L., Rong, Y., Lin, Z., Xing, X., Journal of Materials Chemistry C, 5[4] 2017, 931-936. https://doi.org/10.1039/C6TC03939A

[486] Chen, J., Xing, X.R., Yu, R.B., Liu, G.R., Applied Physics Letters, 87[23] 2005, 231915. https://doi.org/10.1063/1.2140486

[487] Chandra, A., Pandey, D., Mathews, M.D., Tyagi, A.K., Journal of Materials Research, 20[2] 2005, 350-356. https://doi.org/10.1557/JMR.2005.0062

[488] Xu, B., Zu, C.K., Liu, G.Y., Gao, X.P., Zhu, B.J., Yin, X.Y., Journal of Synthetic Crystals, 44, 2015, 159-162.

[489] Chandra, A., Meyer, W.H., Best, A., Hanewald, A., Wegner, G., Macromolecular Materials and Engineering, 292[3] 2007, 295-301. https://doi.org/10.1002/mame.200600422

[490] Sheng, J., Wang, L.D., Li, D., Cao, W.P., Feng, Y., Wang, M., Yang, Z.Y., Zhao, Y., Fei, W.D., Materials and Design, 132, 2017, 442-447. https://doi.org/10.1016/j.matdes.2017.06.061

[491] Kimber, S.A.J., Argyriou, D.N., Yokaichiya, F., Habicht, K., Gerischer, S., Hansen, T., Chatterji, T., Klingeler, R., Hess, C., Behr, G., Kondrat, A., Büchner, B., Physical Review B, 78[14] 2008, 140503. https://doi.org/10.1103/PhysRevB.78.140503

[492] Yiu, Y., Garlea, V.O., McGuire, M.A., Huq, A., Mandrus, D., Nagler, S.E., Physical Review B, 86[5] 2012, 054111. https://doi.org/10.1103/PhysRevB.86.054111

[493] Pomiro, F., Sánchez, R.D., Cuello, G., Maignan, A., Martin, C., Carbonio, R.E., Physical Review B, 94[13] 2016, 134402. https://doi.org/10.1103/PhysRevB.94.134402

[494] Li, Y., Chao, M., Zhang, N., Liang, E., Materials China, 34[7-8] 2015, 509-514.

[495] Li, H., Liu, S., Chen, L., Zhao, J., Chen, B., Wang, Z., Meng, J., Liu, X., RSC Advances, 5[3] 2015, 1801-1807. https://doi.org/10.1039/C4RA08652J

[496] Yamada, I., Tsuchida, K., Ohgushi, K., Hayashi, N., Kim, J., Tsuji, N., Takahashi, R., Matsushita, M., Nishiyama, N., Inoue, T., Irifune, T., Kato, K., Takata, M., Takano, M., Angewandte Chemie, 50[29] 2011, 6579-6582.

https://doi.org/10.1002/anie.201102228

[497] Yamada, I., Shiro, K., Etani, H., Marukawa, S., Hayashi, N., Mizumaki, M., Kusano, Y., Ueda, S., Abe, H., Irifune, T., Inorganic Chemistry, 53[19] 2014, 10563-10569. https://doi.org/10.1021/ic501665c

[498] Yamada, I., Shiro, K., Oka, K., Azuma, M., Irifune, T., Journal of the Ceramic Society of Japan, 121[1418] 2013, 912-914. https://doi.org/10.2109/jcersj2.121.912

[499] Stevens, R., Woodfield, B.F., Boerio-Goates, J., Crawford, M.K., Journal of Chemical Thermodynamics, 36[5] 2004, 349-357. https://doi.org/10.1016/j.jct.2003.12.010

[500] Hosomichi, A., Xue, Y., Naher, S., Hata, F., Kaneko, H., Suzuki, H., Journal of Physics and Chemistry of Solids, 66[8-9] 2005, 1583-1586. https://doi.org/10.1016/j.jpcs.2005.05.062

[501] Yanase, I., Saito, Y., Kobayashi, H., Journal of Thermal Analysis and Calorimetry, 129[2] 2017, 1271–1276. https://doi.org/10.1007/s10973-017-6248-x

[502] Karna, S.K., Li, W.H., Wu, C.M., Wang, C.W., Li, C.Y., Sankar, R., Chou, F.C., Journal of the Physical Society of Japan, 82[9] 2013, 094705. https://doi.org/10.7566/JPSJ.82.094705

[503] Antao, S.M., Acta Crystallographica Section B, 72[2] 2016, 249-262. https://doi.org/10.1107/S205252061600233X

[504] Rebello, A., Neumeier, J.J., Gao, Z., Qi, Y., Ma, Y., Physical Review B, 86[10] 2012, 104303. https://doi.org/10.1103/PhysRevB.86.104303

[505] Hemberger, J., Von Nidda, H.A.K., Tsurkan, V., Loidl, A., Physical Review Letters, 98[14] 2007, 147203. https://doi.org/10.1103/PhysRevLett.98.147203

[506] Vaccari, M., Grisenti, R., Fornasini, P., Rocca, F., Sanson, A., Physical Review B, 75[18] 2007, 184307. https://doi.org/10.1103/PhysRevB.75.184307

[507] Krivovichev, S.V., Filatov, S.K., Burns, P.C., Canadian Mineralogist, 40[4] 2002, 1185-1190. https://doi.org/10.2113/gscanmin.40.4.1185

[508] Novikov, V.V., Matovnikov, A.V., Mitroshenkov, N.V., Shevelkov, A.V., Journal of Alloys and Compounds, 684, 2016, 564-568. https://doi.org/10.1016/j.jallcom.2016.05.235

[509] Jiang, X., Molokeev, M.S., Li, W., Wu, S., Lin, Z., Wu, Y., Chen, C., Journal of Applied Physics, 119[5] 2016, 055901. https://doi.org/10.1063/1.4941266

[510] Novikov, V.V., Mitroshenkov, N.V., Kornev, B.I., Matovnikov, A.V., Journal of Physics and Chemistry of Solids, 104, 2017, 111-116. https://doi.org/10.1016/j.jpcs.2016.12.006

[511] Novikov, V.V., Zhemoedov, N.A., Matovnikov, A.V., Mitroshenkov, N.V., Kuznetsov, S.V., Budko, S.L., Dalton Transactions, 44[36] 2015, 15865-15871. https://doi.org/10.1039/C5DT01406A

[512] Singh, B., Gupta, M.K., Mittal, R., Zbiri, M., Rols, S., Patwe, S.J., Achary, S.N., Schober, H., Tyagi, A.K., Chaplot, S.L., Journal of Applied Physics, 121[8] 2017, 085106. https://doi.org/10.1063/1.4977244

[513] Sheu, G.J., Chen, J.C., Shiu, J.Y., Hu, C., Scripta Materialia, 53[5] 2005, 577-580. https://doi.org/10.1016/j.scriptamat.2005.04.028

[514] García-Moreno, O., Fernández, A., Khainakov, S., Torrecillas, R., Scripta Materialia, 63[2] 2010, 170-173 https://doi.org/10.1016/j.scriptamat.2010.03.047

[515] García-Moreno, O., Fernández, A., Torrecillas, R., Ceramics International, 37[3] 2011, 1079-1083. https://doi.org/10.1016/j.ceramint.2010.11.035

[516] Yao, W., Jiang, X., Huang, R., Li, W., Huang, C., Lin, Z., Li, L., Chen, C., Chemical Communications, 50[88] 2014, 13499-13501. https://doi.org/10.1039/C4CC04879B

[517] Neumeier, J.J., Tomita, T., Debessai, M., Schilling, J.S., Barnes, P.W., Hinks, D.G., Jorgensen, J.D., Physical Review B, 72[22] 2005, 220505. https://doi.org/10.1103/PhysRevB.72.220505

[518] Wang, Y., Wen, T., Park, C., Kenney-Benson, C., Pravica, M., Yang, W., Zhao, Y., Journal of Applied Physics, 119[2] 2016, 025901. https://doi.org/10.1063/1.4940020

[519] Ibid.

[520] Rong, Y., Li, M., Chen, J., Zhou, M., Lin, K., Hu, L., Yuan, W., Duan, W., Deng, J., Xing, X., Physical Chemistry Chemical Physics, 18[8] 2016, 6247-6251. https://doi.org/10.1039/C6CP00011H

[521] Liang, F., Hao, Y., Gao, C., Hu, H., Qin, Y., Journal of the Chinese Rare Earth Society, 30[6] 2012, 703-707.

[522] Gu, C., Yang, Z., Chen, X., Pi, L., Zhang, Y., Journal of Physics - Condensed Matter, 28[18] 2016, 18LT01. https://doi.org/10.1088/0953-8984/28/18/18LT01

[523] Wang, L., Yuan, P.F., Wang, F., Sun, Q., Guo, Z.X., Liang, E.J., Jia, Y., Materials

Chemistry and Physics, 148[1-2] 2014, 214-222.
https://doi.org/10.1016/j.matchemphys.2014.07.037

[524] Abd El All, N., Dalba, G., Diop, D., Fornasini, P., Grisenti, R., Rocca, F., Thiodjio Sendja, B., Vaccari, M., Journal of Physics - Conference Series, 190, 2009, 012066. https://doi.org/10.1088/1742-6596/190/1/012066

[525] Abd El All, N., Dalba, G., Diop, D., Fornasini, P., Grisenti, R., Mathon, O., Rocca, F., Thiodjio Sendja, B., Vaccari, M., Journal of Physics - Condensed Matter, 24[11] 2012, 115403. https://doi.org/10.1088/0953-8984/24/11/115403

[526] Schenk, H., Wolf, M., Mackh, G., Zehnder, U., Ossau, W., Waag, A., Landwehr, G., Journal of Applied Physics, 79[11] 1996, 8704-8711. https://doi.org/10.1063/1.362496

[527] Harunat, K., Ohashit, K., Koiket, T., Journal of Physics C, 19[26] 1986, 5149-5154.

[528] Dmitriev, A.I., Kaminskiĭ, V.M., Lashkarev, G.V., Butorin, P.E., Kovalyuk, Z.D., Ivanov, V.I., Beskrovnyĭ, A.I., Physics of the Solid State, 51[11] 2009, 2342-2346. https://doi.org/10.1134/S1063783409110249

[529] Ishida, I., Journal of the Physical Society of Japan, 39[5] 1975, 1282-1291. https://doi.org/10.1143/JPSJ.39.1282

[530] Bansal, D., Hong, J., Li, C.W., May, A.F., Porter, W., Hu, M.Y., Abernathy, D.L., Delaire, O., Physical Review B, 94[5] 2016, 054307. https://doi.org/10.1103/PhysRevB.94.054307

[531] Liu, G., Zhou, J., Wang, H., Physical Chemistry Chemical Physics, 19[23] 2017, 15187-15193. https://doi.org/10.1039/C7CP00815E

[532] Seyidov, M.Y., Suleymanov, R.A., Journal of Applied Physics, 108[6] 2010, 063540. https://doi.org/10.1063/1.3486211

[533] Mammadov, T.G., Abdullayev, N.A., Seyidov, M.Y., Suleymanov, R.A., Yakar, E., Japanese Journal of Applied Physics, 50[5-3] 2011, 05FD06.

[534] Abdullayev, N.A., Mammadov, T.G., Suleymanov, R.A., Physica Status Solidi B, 242[5] 2005, 983-989. https://doi.org/10.1002/pssb.200402126

[535] Chen, X.L., Yang, Z.R., Tong, W., Huang, Z.H., Zhang, L., Zhang, S.L., Song, W.H., Pi, L., Sun, Y.P., Tian, M.L., Zhang, Y.H., Journal of Applied Physics, 115[8] 2014, 083916. https://doi.org/10.1063/1.4867217

[536] Porter, R.S., Weeks, N.E., Capiati, N.J., Krzewki, R.J., Journal of Thermal

Analysis, 8[3] 1975, 547-555. https://doi.org/10.1007/BF01910133

[537] Chen, F.C., Choy, C.L., Young, K., Journal of Polymer Science A2, 18[12] 1980, 2313-2322.

[538] Stobbe, R.A., Hägele, P.C., Journal of Polymer Science B, 34[5] 1996, 975-980. https://doi.org/10.1002/(SICI)1099-0488(19960415)34:5<975::AID-POLB15>3.0.CO;2-8

[539] Baughman, R.H., The Journal of Chemical Physics, 58[7] 1973, 2976-2983. https://doi.org/10.1063/1.1679607

[540] Choy, C.L., Chen, F.C., Young, K., Journal of Polymer Science A2, 19[2] 1981, 335-352.

[541] Wand, C.R., Bolton, K., Journal of Polymer Science B, 54[21] 2016, 2223-2232. https://doi.org/10.1002/polb.24131

[542] Hua, Y., Ni, Q.Q., Yamanaka, A., Teramoto, Y., Natsuki, T., Advanced Composite Materials, 20[5] 2011, 463-475. https://doi.org/10.1163/092430411X576774

[543] Takezawa, A., Kobashi, M., Kitamura, M., APL Materials, 3[7] 2015, 076103. https://doi.org/10.1063/1.4926759

[544] Shen, X., Viney, C., Johnson, E.R., Wang, C., Lu, J.Q., Nature Chemistry, 5[12] 2013, 1035-1041. https://doi.org/10.1038/nchem.1780

[545] Landert, M., Kelly, A., Stearn, R.J., Hine, P.J., Journal of Materials Science, 39[11] 2004, 3563-3567. https://doi.org/10.1023/B:JMSC.0000030707.91634.5f

[546] Szafrański, M., Journal of Materials Chemistry C, 1[47] 2013, 7904-7913. https://doi.org/10.1039/c3tc31609b

[547] Jones, R.H., Knight, K.S., Marshall, W.G., Clews, J., Darton, R.J., Pyatt, D., Coles, S.J., Horton, P.N., Cryst Eng Comm, 16[2] 2014, 237-243. https://doi.org/10.1039/C3CE41909F

[548] Nicolaï, B., Rietveld, I.B., Barrio, M., Mahé, N., Tamarit, J.L., Céolin, R., Guéchot, C., Teulon, J.M., Structural Chemistry, 24[1] 2013, 279-283. https://doi.org/10.1007/s11224-012-0078-z

[549] Bhattacharya, S., Saha, B.K., Crystal Growth and Design, 12[10] 2012, 4716-4719. https://doi.org/10.1021/cg300980s

[550] Haas, S., Batlogg, B., Besnard, C., Schiltz, M., Kloc, C., Siegrist, T., Physical Review B, 76[20] 2007, 205203. https://doi.org/10.1103/PhysRevB.76.205203

Keywords

www.ingramcontent.com/pod-product-compliance
Lightning Source LLC
Chambersburg PA
CBHW071232210326
41597CB00016B/2024